EARTHSHOT

EARTHSHOT

HOW TO SAVE OUR PLANET

COLIN BUTFIELD AND JONNIE HUGHES

WITH FRED PEARCE

JOHN MURRAY

First published in Great Britain in 2021 by John Murray (Publishers)
An Hachette UK company

1

Copyright © Colin Butfield and Jonnie Hughes 2021

Introduction © HRH The Duke of Cambridge, 2021

A CIP catalogue record for this title is
available from the British Library

The acknowledgements on page 338 constitute an extension
of this copyright page.

Hardback ISBN 978 1 529 38862 6
Trade Paperback ISBN 978 1 529 38863 3
eBook ISBN 978 1 529 38865 7

Typeset in Sabon MT Std by Palimpsest Book Production Ltd,
Falkirk, Stirlingshire

Printed and bound in the United States of America

John Murray policy is to use papers that are natural, renewable
and recyclable products and made from wood grown in sustainable forests.
The logging and manufacturing processes are expected to conform to
the environmental regulations of the country of origin.

John Murray (Publishers)
Carmelite House
50 Victoria Embankment
London EC4Y 0DZ

www.johnmurraypress.co.uk

Contents

Introduction

The sight that greets you on arrival at the Hoanib Valley in the Kunene region of Namibia is breath-taking. The desert landscape is rocky and bare, criss-crossed with ancient, dry riverbeds that are now used as roads, and punctured by the odd tree and bit of scrub. The setting sun behind the mountains makes you stop in your tracks. This place has a majestic otherworldly beauty.

What makes the environment even more breath-taking is the wildlife that now thrives here. Herds of springbok and oryx pick their way through the dusty plains. Desert-adapted giraffes amble along with an elegant gait. Elephants shelter from the beating sun under the shade of large trees. And if you are willing to get up early, be patient, and have a bit of luck on your side, you might spot one of the region's precious, free-ranging black rhino or even a desert lion.

It was here, in September 2018, under the vast Namibian desert sky, that The Earthshot Prize was born.

I had wanted to visit this region for years, and it didn't disappoint. But beyond the visual wonders was an even more inspiring story. Namibia is a leader in developing and implementing a modern way of doing wildlife conservation that puts people first. In the 1980s, faced with the rampant poaching sadly seen across much of the world, the government granted communities the right to create conservancies: areas with defined borders and a management structure outside of the national parks. The idea was to give communities the right to manage and benefit from the wildlife they live beside.

Since 1998, Namibia has created eighty-six such conservancies, covering nearly 20 per cent of the country and 9 per cent of its population. As a result, Namibia's elephant population has more than doubled and black rhinos – once near extinction – have grown to be the largest free-ranging population in the world. Rural economies have been enhanced and diversified, bringing new jobs, tourism and investment. Tourism accounts for just over 11 per cent of Namibia's GDP. Covid-19 has of course made life very hard, but across the board, local communities' commitment to the long-term benefits of the conservancy model remains strong.

You might be wondering how I went from a 5 a.m. start to catch a fleeting glimpse of a shy black rhino in the north-west corner of Namibia, to building a team to deliver the most ambitious environmental prize in history. The answer lies in a crucial disconnect this visit clarified for me, between the optimism and determination I saw on the ground, and the despair and anger that would come to dominate headlines just a few weeks later.

The rich wildlife that I saw thriving on that visit struck a real chord. The community conservancy model is a prime example of how a simple, positive solution can have wide-reaching benefits for both humans and nature. Most importantly of all, it is a success story that can be replicated and scaled. I wanted to find a way to bottle that innovation and community spirit and mass-produce it globally.

But when I returned to the UK, just as the world was gathering again for the next round of climate change negotiations in Poland, I was hit by a wave of global pessimism. The headlines were dominated by a sense that world leaders were not moving fast enough. There was widespread finger pointing and political and geographical division. To those of us following at home, it wasn't an inspiring sight.

I understood why the mood was full of despair, of course. The challenge facing our planet is immense. We were about to enter what scientists say is the most consequential decade in history. Humans have taken too many fish from the sea. We have cleared too many trees, burnt too much fossil fuel, and produced too much waste. The damage we are doing is no longer incremental but exponential, and we are fast reaching a tipping point.

The science tells us that if we do not act to restore our planet by 2030, the damage will be irreversible, and the effects will be felt not just by future generations but by all of us alive today. What is more, this damage will not be felt equally by everyone. It is the most vulnerable, those with the fewest resources, and those who have done the least to cause climate change, who will be impacted the most.

The facts look terrifying, and I could see that this risked making people feel like they might as well give up. The global debate felt *too* complex, *too* negative, *too* overwhelming. It seemed to me, and this is backed up by my team's research, that there was a real risk that people would switch off; that they would feel so despondent, so fearful and so powerless, that any real hope of progress would come to a halt. You could summarise this mood with a simple equation: **urgency + pessimism = despondency.**

This despondency also jars with my own experiences, and those that inspired my grandfather and father to be pioneers in the environmental movement. Following in their footsteps, I have seen people all over the world face what seem like insurmountable challenges yet come together with collective ambition, and a can-do-spirit, to find solutions to them. I strongly believe that change is possible, when you put your mind to it. I started thinking about what to do to change the equation to something else: **urgency + optimism = action.**

The most famous example of using optimism to rise to a great challenge is the Moonshot. Standing in front of a large crowd at Rice Stadium in Houston, Texas, in September 1961 President John F. Kennedy audaciously declared: 'We choose to go to the moon in this decade and do the other things, not because they are easy, but because they are hard; because that goal will serve to organize and measure the best of our energies and skills, because that challenge is one that we are willing to accept, one we are unwilling to postpone.' Kennedy's Moonshot was a dream so ambitious it required unparalleled innovation and effort from a huge multinational team.

When the Moonshot *was* accomplished, *within* the decade, it was a defining achievement in our global history.

The incentives were mixed, of course – the space race was a key component of the Cold War. But it was an incredible demonstration of our talent for making the impossible possible. And, crucially, the technological advances brought about by the Apollo space programme still bring benefits to this day. They inspired developments in lightweight materials, heart monitors, CAT scanners, breathing equipment and solar panels.

So, at the end of 2018, inspired by the brilliant solutions I had seen on the ground in Namibia and elsewhere, and at the same time horrified by the cliff edge the scientists were predicting, yet determined not to give up, I set about asking how I could play a helpful role in bridging that disconnect. I wanted to recapture Kennedy's Moonshot spirit of human ingenuity, purpose and optimism, and turn it with laser-sharp focus and urgency on to the most pressing challenge of our time – repairing our planet.

In the months of scoping that followed, my team and I spoke to people around the world – from activists to scientists, business leaders to prime ministers, and conservationists to filmmakers. I pushed them hard, because I wanted to make sure that whatever I did was collaborative and complementary, had widespread support, and would have the impact that was needed.

Together we crafted an idea to replicate the Moonshot for our generation's challenge. Working with an incredible range of partners and experts, we identified five great underpinning goals that, if they are met, will offer us the greatest chance of a stable and thriving future.

We called these five great challenges, the **Earthshots**. They are:

- Protect and Restore Nature
- Clean Our Air
- Revive Our Oceans
- Build a Waste-Free World
- Fix Our Climate

Each of them will need a global, collaborative effort to achieve, but once achieved, will help to put our planet back on course again. There were a few moments along the way when the whole idea seemed a bit daunting. What happens if we fail? But of course, failure is not an option. We simply have no choice but to succeed, and I want to do all I can to contribute. We all owe it to the generations that follow us, our children, and their children, to be outlandish in our ambition and stubborn in our optimism.

Having set those Earthshot goals, I committed to using the unique position that I have to celebrate, reward and support those who are doing remarkable things, to help them achieve these things across every sector of society and in every part of the world. The result is The Earthshot Prize.

Deciding to create a prize, as opposed to any other form of initiative, was again a lesson from history. When facing a big challenge that takes many years to achieve, it is easy to lose momentum. We need to see progress, be inspired by achievements, witness leadership and courage against the odds. As a species we often define ourselves by collective stories whose narratives shape our cultures and values. Achieving these Earthshots will need regular chapters of an ever-evolving story. It will need heroes, revelations and hope.

Over the next decade The Earthshot Prize will seek out, celebrate and support fifty inspiring solutions. They must be game-changing, offer a significant impact for human-kind, and have potential to be scaled globally. Five inspirational winners a year for ten years.

Winners of the Prize will receive not only global public recognition and a reward of £1 million but also the support they need to realise their ideas at scale. Over the next ten years we want the fifty winners to be acclaimed, like the returning Apollo astronauts were, as heroes for our time and leaders of the greatest endeavour of our generation.

The Earthshot Prize is about much more than awarding achievement. It is about revealing that an aspirational, sustainable future is ours to win, if we are willing to reach for it. It is about building a global coalition of partners, to inspire entrepreneurs, activists, innovators and commu-nity leaders to reach for the stars and to incubate and grow the sustainable solutions that will change the trajectory of this planet.

The global response to the Covid-19 pandemic is evidence that all this *is possible*. The funds flowing into the recovery effort demonstrate how much can be achieved when those in positions of power come together and decide to act. We've built hospitals overnight, repurposed factories and poured billions into the search for a vaccine and better treatments. And we've been inspired by heroes emerging in every community across the world.

Young people no longer believe that change is too difficult. They've witnessed the world being turned on its head. They believe that the climate crisis and the threat to our biodiversity deserves our full attention and ambition. And they're right. So now is the time for each one of us to show leadership.

I know it will take more than fifty solutions to solve the five Earthshots. But over the next decade we hope to inspire people in all corners of the world, from all sectors of society, to do their bit too. Whether you're a farmer, a tech entrepreneur, a politician, a banker, a fisherman, a community leader, a mayor or a student. Every single one of us has a role to play in harnessing whatever opportunity we have. This is a global team effort.

Those of us alive today have been handed the momentous task of determining the fate of our descendants for centuries to come. We have the power to choose to react in time and bring about a better future. That is an honour, not a burden.

HRH Prince William, 2021

PART 1

THE CRUCIAL DECADE

1

Ten Years to Save the World

After more than 200,000 years of human history, our species stands at a critical moment. The choices we make in the next ten years will determine our future. In this decade we have a final opportunity to choose between a future where we thrive on a healthy, stable planet in balance with the natural world, or a future where we continue on our current path as often unwitting yet rapacious agents of its destruction – and of ours.

Our world is full of stories of how we have destabilised our planet and its life support systems – the stable climate, reliable rainfall, beach-lined shores, rich soils and self-cleansing air that have successfully sustained ever greater numbers of us. We know this destruction cannot go on. To change direction, we need to face the truth. But the stories of doom can sometimes so overwhelm our thinking that we can become defeatist. We cannot see a way ahead. We come to believe that little can be done. Nothing could be farther from the truth.

So we badly need to acknowledge not just the urgency

of the crisis we are in, but also that all around us is evidence of natural recovery and human redemption. We need inspiring tales of how we can undo the harm and set nature and its vital systems on a path to revival. This book will look at the dark and destructive side of our relationship with our world: we have to understand that in order to appreciate the urgency and know what to do better. It will pull no punches. But more importantly, it will explore the good side of humanity's impact on our planet. We will encounter stories to inspire us, provide optimism and put us on the path we know we must take, in what scientists say will be our defining decade as a species. These will be human stories. They will show us that it's not OK to give up or give in just because our challenges look complicated or intractable. They will show us that people can fight and work for change and win. Because so often transformation starts with the inspiring example of one human being.

Here is one hero of ours: Juan Castro, the saviour of coral reefs around his home fishing village of Cabo Pulmo, in the Gulf of Cortez on the Pacific coast of Mexico. Arguably too, the man who started a revolution to save the world's oceans. A generation ago, his fishing community was dying. Just offshore, the coral reef that had nurtured the fish caught in their nets had been wrecked, both by weighted gill nets spread by big trawlers sailing in from outside the gulf, and by ships' anchors knocking the coral into small pieces. As the reef crumbled away, the fish departed.

The coral reefs of the Gulf of Cortez were once famous.

They were the most northerly reefs on the Pacific east coast, and probably the oldest, at 20,000 years. When the American author John Steinbeck visited in 1940, he described how the reef 'pulsed with life, little crabs and worms and snails'. He wrote: 'One small piece of coral might conceal 30 to 40 species, and the colours on the reef are electric.' The famed French marine explorer Jacques Cousteau called the gulf 'the world's aquarium'. Even when Juan was a young lad out fishing with his father, there were shoals of sharks in the reef. To capture one of them on a line from a small boat was great sport and lucrative business.

But then, as the coral crumbled, the sharks went away, along with most of the other fish. Cabo Pulmo looked likely to disappear from the map, too, with its driftwood shacks washed back into the ocean. One day, Juan said years later, he had a revelation while diving among the coral with his father. He saw the beauty of the reef despite the scars left by the anchors and nets. He decided to fight back. He found a local marine professor to research the reef, and they began a campaign to save the reef and its life.

The professor suggested that the villagers should stop fishing on the few surviving stretches of coral where shoals remained, and they should press the government to keep out foreign vessels and turn the reefs into a marine protected area. It sounded an impossible ask. The villagers would have to voluntarily give up what remained of their main source of food and income. It seemed like Juan and his community were being punished for having one of the few reefs left that was worth saving. But the professor promised that if the reef was allowed to recover, it could again become a breeding ground for fish. And thanks in large part to the persuasive powers of Juan, the village agreed to the creation of a no-fishing zone, and the protected area was created in 1995.

For almost ten years they waited and watched. It was hard. Proud fishermen were reduced to shopping for food in markets, with government-issued food vouchers. They could see the shoals of fish returning in the crystal blue waters, swimming through coral that was growing iridescent once again. They were sorely tempted to break the fishing ban. Perhaps a few did. But after ten years there was a red-letter day. The first sharks came back to Cabo Pulmo. As top predators, they were the proof that the reef's ecological virility – its ability to provide fish for the coastal communities on the Gulf of Cortez – had returned. Now there are groupers and snappers and eels and even jack tuna.

The marine biologists were back too, counting and calibrating. They said fish numbers had increased by more than 400 per cent in the no-fish zone. Better still, the fish nurtured there were beginning to repopulate other nearby reefs. And besides the big fish and glistening healthy coral, grey whales returned to their old calving grounds just offshore. The biologists were excited because this was one of the most dramatic turn-arounds ever witnessed in a coastal fishery, and proof that marine protected areas could kick-start ecosystem revival in double-quick time.

The fishermen were back in business, too. Local laws

inside what was now a national park allowed the villagers to set their nets, but kept out foreign trawlers. Soon, to avoid relying on the fish alone, the villagers were inviting in tourists. They set up seafood restaurants, guesthouses and shops selling diving gear.

A decade of conservation had brought short-term pain, but ultimately much greater gain. By nurturing their reef's recovery, the people of Cabo Pulmo have restored an ecosystem that can again deliver nature's bounty – for fishers and for tourists.

Juan Castro's niece Judith is a proud spokesperson for her community and can see a bigger picture. Her fellow villagers made 'a leap of faith' and it paid off. 'We need to have Cabo Pulmos all over the world,' she says. 'It is precisely what the entire planet needs. It is an example of what can still be done for the planet.' Marine biologists say that Juan's galvanising of his fishing community in the name of conservation was one of the starting points for what has become a global movement to protect marine ecosystems and restore fisheries inside protected areas. It provided some of the first and best evidence that if you stop fishing, the fish will come back – quickly and in profusion, restocking surrounding areas of ocean. 'Species come back quickly – in three or five or ten years,' says Boris Worm of Dalhousie University in Nova Scotia, Canada. 'And where this is done we see immediate economic benefits.'

Juan's story shows that, given the chance, even in the most difficult circumstances nature will rebound, will restore itself. All we have to do is give it the chance. We don't have to cut ourselves off from nature to save it. When environmentalists use terms like 'sustainability', this is what they mean – a way of living that can last forever.

. . .

Not many people have heard of Juan Castro. That is a shame. But Kenyan tree lover Wangari Maathai is rightly a hero to millions. She gave up her career as an academic to spend most of her life on a mission to inspire millions of poor rural women to plant trees across their deforested country, and to force sceptical politicians to put communities in charge of the forests. Before her death, in 2011, she had established 6,000 community nurseries – all run by women – that had planted 30 million trees on farms, in gardens, at the roadside, outside schools and public buildings, and even in forests. For her temerity in demanding that the government of President Daniel arap Moi join her in the task, she was repeatedly beaten up, subjected to death threats and for a while had to go into hiding.

But she won. After Moi was ousted from office, she briefly became an environment minister, and passed into law in 2005 a Forest Act that created more than 300 democratically-elected Community Forest Associations. They gave local people control over their local forests, allowing them to graze their livestock, cut firewood and construct beehives, so long as they did not plough the land or build their homes among the trees. Maathai's law also gave protection to the country's forested mountain areas – the 'water towers' that maintained the nation's river flows and water supplies.

Her legacy lives on. Years later, at Kimunye village near Mount Kenya, Sarah Karungari tends a dozen beehives in a clearing on the edge of one of the water-tower forests. 'Maathai changed everything here,' she says. Before her law, forest rangers would have torn down her hives and prosecuted her for invading the forest. Now, they recognise that villagers such as her can be both users and protectors of the forests. 'People who used to be poachers and illegal loggers are now defending the forests,' Simon Gitau, senior warden for the forests on Mount Kenya, agrees. 'Farming communities know their ecosystems better than outsiders. We have to work with them if we want to protect the forests.'

Maathai was a hero within the international environment community for many years. Among feminists too, she was feted for her courage in the male-dominated world of Kenyan politics. Then, in 2004, she was awarded the Nobel Peace Prize for her work. The prize committee said: 'Peace on earth depends on our ability to secure our living environment. Maathai stands at the front of the fight to promote ecologically viable social, economic and cultural development in Kenya and in Africa . . . She thinks globally and acts locally.' Her commitment to her local community made her globally famous.

The Kenya she left behind is a country more forested than for decades. Her Green Belt Movement is now run by her daughter, Wanjari. It has assisted in the planting of more than 50 million trees. An estimated one-third of Kenya's trees are located on farms and within communities. Many grow on the increasingly tree-covered mountain 'water towers' that Maathai first identified. On the Aberdare mountain range – the source of four of Kenya's seven largest rivers, including the Tana River, which fills taps in the Kenyan capital, Nairobi – the Green Belt Movement is educating farmers to conserve trees on their land. Forest cover has increased there by a fifth since 2005.

When Maathai began her mission to restore Kenya's trees back in 1977, her country had the one of the highest birth rates in the world, and had ground to make up in pursuing higher living standards for its people. Bringing back forests in such a fast-growing and industrialising country seemed an impossibility, even to most environmentalists. She changed that perception. And Kenya is not alone. Like Juan Castro and his marine protected area, Maathai's insight and enthusiasm has changed our views about how nature can be restored.

• • •

There are heroes all over the world. Back in 2006 a young British farming student and part-time gardener called Rob Hopkins moved to Totnes, a small market town in southwest England. Inspired by a desire to break away from reliance on global food markets, he set up a community food-growing project in the town, to encourage self-sufficiency. The idea took off – in Totnes and far beyond. In two years, there were local groups in Ireland, Scotland, Italy, Japan, Spain and Sweden. They called themselves 'transition towns', and the following year new initiatives were set up in Germany, Australia, Denmark and France.

Hopkins didn't want a big organisation or to set up a political movement. There was no blueprint other than enthusiasm and localism. Like his vegetables, he wanted organic growth. He wanted people to copy each other in doing things for themselves, and to set a social fire burning. Still, he oversaw a growing network of transition towns, devoted to promoting everything from local food businesses to energy companies, social enterprises and even local currencies by, as he puts it, 'sparking entrepreneurship, reimagining work, reskilling themselves and weaving webs of connection and support'. It works, he says, because it combines 'heart, head and hands' to bring communities together to save their worlds.

The Transition Network now has thousands of communities doing their own thing in more than fifty countries, from a free store in Pennsylvania to harvesting the rain for favelas in São Paulo, from a repair cafe in Pasadena to permaculture in Peterborough, Canada, and from zero-carbon neighbourhoods around Bologna to community gardens in Brussels. All are different. That is what it means to be local. But whatever they are doing, they feel optimistic. They believe they are the future. You've probably never heard of Rob Hopkins before. He probably rather likes it that way. But he is a hero, nonetheless.

• • •

Our planet is in crisis. An ecological crisis. Lost forests, parched soils, dammed and poisoned rivers, empty oceans, filthy air and compromised climate leave it in peril. Since this is our only home, the predicament leaves us humans in peril, too. Our tragedy is to be master of all we survey. Many scientists say we have only a decade to pull back from the brink, to head off runaway effects of our mastery. But those same scientists also say there is hope. We can get out of this fix. We have the tools. What we need above all is belief. To get rid of the excuse that nothing can be done. Because it can. The three heroes above are not making futile gestures; they are beating a path out of our crisis. A path that we, our governments and our entire societies must follow.

This book attempts to reinforce that hope, and to chart some of the route back to stability and balance. Much of what is needed is abundantly clear. We need to break the practices that make everyday waste seem normal and create 'circular' systems that reuse everything. We need to keep carbon in the ground, restore nature, and cleanse our air and oceans – the great shared domains that some call the 'global commons'.

But there is no simple blueprint for achieving these ends. Human ingenuity, inventiveness and resourcefulness must come to our aid. The good news is that humanity has always had such skills in abundance. We are great problem solvers. That attribute may have got us into this mess, by coming up with all manner of ways in which we can dominate and devour our world. But it can also help us in our new quest to live in harmony with that world.

Scientists invented a new word at the start of the twenty-first century to describe our new relationship with the world. They say we are now in the 'Anthropocene'. It is a new geological era. After the Pliocene, Pleistocene and Holocene, we are now in the human epoch, where humans are the major influence on how the planet works. The problem is that while we are now in charge, we are not yet in control. For many, that means the Anthropocene can only end in tears. But it needn't be like that. We can have a 'good' Anthropocene. But we have to be the ones to make it happen.

• • •

How did we get here? Humanity's dominance over planet Earth did not happen suddenly. Humans have been a growing presence for a long time. Our species, *Homo sapiens*, has been around for more than 200,000 years. Through the last two ice ages we gradually saw off predecessors and co-inhabitants such as the Neanderthals. At some points our numbers dropped to just a few thousand, but we endured. Whether we did so because we were cleverer or luckier is unclear. But we emerged from the end of the last ice age, around 12,000 years ago, as the last human species standing.

As the ice that once extended across much of the land surface of the northern hemisphere – as far south as the River Thames in England – retreated, our small populations grew and colonised most of the planet. The forests and grasslands, marshes and mountains and coastal plains, all found themselves occupied by humans. We began as hunters of animals and gatherers of nature's fruits. But, as the climate stabilised, rainfall became more predictable and we grew more ambitious. We began planting our favourite fruit trees, cultivating crops and domesticating animals for meat, milk and leather, and to do some of the hard work for us. We began cutting and burning clearings in forests to grow our crops.

All this started to transform how our world looked. We sometimes think that until recently nature was pristine and intact, barely touched by humans. But the more archaeologists dig into our past, the less things look like that. They recently uncovered an area on the shore of Lake Malawi in East Africa that was permanently deforested when humans set fire to it more than 90,000 years ago, apparently deliberately. 'This is the earliest evidence of humans fundamentally transforming their ecosystem with fire,' says Jessica Thompson from Yale University.

With time, such activities became widespread. Geographer Erle Ellis of the University of Maryland, Baltimore County reckons that soon after the end of the ice age 'nearly three-quarters of land on Earth was inhabited, used and shaped by people. Areas untouched by people were almost as rare then as they are today.'

Of course, our touch was lighter back then – our global population was likely as low as 4 million and each of their demands far smaller than most of ours today. Mostly, we did not permanently destroy forests. We created clearings to cultivate crops for a few years, before planting some trees and moving on to let the forest recover. Often, far from wrecking ecosystems, this form of shifting cultivation increased the number of plant species by spreading seeds and our favourite trees. Our ancestors often improved soils too, by burying their household waste. They were keen recyclers. Archaeologists digging into many tropical forest soils have found rich mixes of discarded food, excrement and even charcoal that all continue to nurture soils thousands of years on. Such enriched soils are so frequent in the Brazilian Amazon that researchers call them *terra preta*, which is Portuguese for 'black soils'.

As our numbers grew our impact on the land increased. We moved from living in villages to creating cities, where crafts and learning flourished. Growing links between communities, and skills in passing on knowledge from one generation to the next, enabled us to spread knowledge and organise – in particular in the development of farming systems to feed growing populations. Many of the earliest civilisations were in arid regions that required irrigation networks – tapping the waters of the Indus in modern-day Pakistan, the Yellow River in China, the Tigris and Euphrates in Iraq and the Nile in Egypt. We herded cattle and smelted iron, chopping down forests to do so. We created transport networks on land and at sea, and began trading across continents and over the oceans in metals and silk, spices and enslaved people. But things really kicked off with the industrial revolution, which got under way in Britain just over two centuries ago. For the first time we began burning coal on a large scale, to power manufacturing machinery and railways, and later to generate electricity to power cities. Horse power was replaced by steam power. Then in the twentieth century we began tapping oil, and the age of the car roared into view. The consequence of all this technology was that humankind was no longer constrained by the work

15

rate of their own or their animals' muscles. We were discovering ways to do more work, extracting more materials to make more stuff, to transport more products and to intensify how we used the land.

We developed new industrial processes, perhaps most importantly the Haber-Bosch process that converted nitrogen gas in the air into nitrogen fertiliser. That massively increased how much food we could grow. Meanwhile, scientists found ways to fight the diseases that once killed most children and cut short the lives of adults. Soap, sewers, vaccines and antibiotics meant that by the late twentieth century – and probably for the first time in human history – most kids got to grow up, rather than dying in infancy. There was food to feed most of them. Life expectancy rose from thirty or forty years to seventy or eighty years, first in Europe and North America, but increasingly across the world.

These breakthroughs triggered a demographic explosion. The human population grew exponentially. From 1 billion at the start of the nineteenth century to 2 billion in the 1920s, 3 billion in 1960, 4 billion in 1975, 5 billion in 1987, 6 billion in 1999, 7 billion in 2013, and likely 8 billion before the 2020s are out. Although population continues to rise, the boom is probably over; though we will likely have 10 billion people by the end of the century.

In the twentieth century our consumption of natural resources – from metals and fossil fuels to forests and fish – grew even faster than our population. In some rich countries that surge of ever-rising demand may be largely over, although it is continuing at a very high level, far above poorer nations. As we shall see later, some richer societies may even be starting to 'dematerialise'. But countries with lower consumption levels will increase theirs as they strive for a good standard of living.

Take China, as one of the most startling examples. Thirty years ago, Beijing was a city of bicycles. They queued by the thousands at traffic lights on roads where cars were rare. Today, it is the bicycles that are rare. Six million cars swirl around eight ring roads encircling the capital of the world's most populous nation. The metropolis chokes in smog for much of the year. The old *hutong* pedestrian neighbourhoods have been replaced by high-rise apartments and shopping malls with underground car parks.

The industrialisation of China has propelled a quarter-billion people from dirt-poor rural villages to modern megacities. The country now has more cars on its roads than America does. Two-thirds of all the skyscrapers erected in the world in 2016 went up here. China's demands of breakneck construction and galloping consumption are transforming the planet as a whole. The country has recently been consuming more cement every three years than the United States managed during the entire twentieth century. It now consumes 60 per cent of the world's cement, 50 per cent of its iron ore and coal, and 40 per cent of its lead, zinc and aluminium.

Domestic Chinese demand is growing. Its retail market is now, like its car fleet, bigger than America's. But China has the world's largest population, and the average citizen still has a lower standard of living and environmental impact than the average European or North American. The country's extraordinary appetite for the materials of the Earth in large part reflects the rest of the world's enjoyment of stuff made in China. The country is now often called the workshop of the world, a phrase that applied to Britain in the nineteenth century. It produces three-quarters of our air conditioners, two-thirds of our mobile phones and photocopiers, and one-third of our cars and televisions. The West has sent much of its own environmental footprint overseas, masking its true impact.

China is only following the path to consumerism adopted by Europe and North America many decades ago. But many other countries are following in its wake. And the consequences of so many people requiring so many things are profound. The pioneer researcher on the human 'population bomb', California biologist Paul Ehrlich, noted that our impact on the planet is a combination of our numbers, our consumption, and how we satisfy that consumption. Our numbers are slowly starting to stabilise, and with greater global equality may do so faster, but our consumption demands continue to soar. Our hope is that the third element in his equation may come to our aid. Certainly, over the next decade, addressing it is our biggest and quickest route to reducing humanity's footprint on the planet, and to finding ways in which 10 billion of us can live well on a planet that once had to cater for just a few thousand of us. But it requires a fundamental rethink of how we live.

• • •

Our ancestors consumed few natural resources and had little waste because most of the things they consumed were the fruits of nature, and nature constantly recycles everything. What they took mostly regrew. But our modern world has been very different. The way we generate energy to power our world looks like our worst habit. Coal and oil, along with the natural gas that many of us burn to heat our homes, are all made of carbon. They are often called fossil fuels because that is what they are: the fossilised remains of ancient trees and marine organisms that were buried beneath the earth or on the seabed hundreds of millions of years ago – before even our greatest predecessor the dinosaurs ruled the Earth.

Burning this carbon fuel generates most of our energy. It makes heat to drive industrial processes or steam to drive turbines for electricity. It powers pistons to transport us around. But the burning of this carbon creates billions of tonnes of carbon dioxide each year, which is released into the atmosphere. And as we have all become aware in recent times, carbon dioxide warms the planet. It is our atmosphere's thermostat. We need some carbon dioxide in the air or our world would be frozen; but too much of it and we start to overheat. Nature used to maintain stable levels, but our emissions are resetting the thermostat. This 'greenhouse effect', which almost nobody had heard of half a century ago, is now probably the number one threat to our future, and ultimately to life on Earth.

As we change the climate, we are also transforming the surface of our planet, replacing nature with concrete and steel, bricks and asphalt. Scattered across the globe today are more than a trillion tonnes of these materials. Most is on land, but we are invading the oceans, too. Since 2000 a staggering 1 million kilometres of fibre-optic cables, enough to go round the Earth twenty times, have been laid on the seabed to carry most international digital telecommunications.

Ron Milo of the Weizmann Institute of Science, in Israel, calculates that we make 30 billion tonnes of materials for our use every year. At the apex is concrete. With 10 billion tonnes more in circulation annually, it is the most widely used substance on the planet, excepting only water. The accumulated pile of our materials has been doubling every twenty years for more than a century now. One way or another, the average citizen on the planet adds to it by more than our own body weight each week.

A separate study found that London's 9 million inhabitants are collectively responsible for some 40 million tonnes of construction materials a year, plus another 8 million tonnes of paper, plastic, food, glass and metals. That's around 100 kilograms for each Londoner every week. Not all this is for personal use. Most of the newly accumulating material is in buildings and infrastructure such as highways, bridges and sea walls. Still, all this material obliterates nature – entombing soils, barricading rivers, draining wetlands and breaking ecosystems into tiny fragments.

Extracting these materials from the earth is equally destructive. One of the least discussed but most pervasive extractive industries is sand mining. Sand is the main ingredient in concrete, asphalt and other building materials, as well as for making glass. We currently use more than 40 billion tonnes of sand each year – five times more than our demand for coal. Digging it up is the world's largest mining business. You would think we could get all we need from the world's deserts. But it turns out that the grains in desert sand have mostly been rounded by wind erosion and do not bind well in concrete and other materials. So most sand comes from excavating riverbeds, deltas and coastlines – usually far faster than rivers or tides can replace what is lost.

The Chinese megacity of Shanghai has been built largely from sand dug from the Yangtze River, wrecking its ecosystem. The islands state of Singapore has grown its land area by 20 per cent by importing half a billion tonnes of sand dug from the coastlines of neighbouring Asian countries. Cambodia's shoreline is retreating as a result. In the United States, sand mining in Houston is said to have caused sedimentation that played a part in the flooding during Hurricane Harvey in 2017. Dubai's construction boom has taken even more sand – a strange case of sending sands to Arabia. In India there are sand mafias cashing in on a construction boom.

Maybe even more staggering is that our accumulated trillion tonnes of inanimate stuff is greater than the planet's entire living biomass of trees, plants and animals. At the start of the twentieth century human-made materials were equivalent to about 3 per cent of living matter on the planet. Now the figure is more than 100 per cent. In this way, besides covering the planet with built infrastructure, we have in the past three centuries been wrecking nature's own infrastructure. Six trillion trees have become 3 trillion trees, for instance. Once, we lived in what scientists called a 'biosphere'. Now we live in a

21

'stuffosphere'. We have almost literally turned the planet into a concrete jungle.

And what life remains looks unlike anything before. More than 40 per cent of the ice-free land surface of the Earth has been taken over for agriculture. Only 4 per cent of the total mass of mammals left is wild animals – from shrews to blue whales. Humans make up 36 per cent of all mammals, and our cattle, sheep, goats and other domesticated livestock account for the remaining 60 per cent. At any one time, there are more than 20 billion domesticated chickens on the planet. They weigh more than all the world's wild birds put together.

It is a bleak story of human avarice and short-sightedness. We are only now understanding the extent of our abuse of the planet and realising in the nick of time that we are reaching a critical point. The next ten years will be the most important in human history – our last chance to change course. But to meet the challenge we need to retain some virtues from the past. We should remember that, during the same period when we have blighted the planet, our problem-solving ingenuity cured diseases, gave warmth, light and safety to billions, and enabled nations to choose to cooperate for the greater good of all. And we should remember that even at this moment of planetary crisis, there are seeds of survival and resurgence. Nature is resilient; humans are innovative. We can fight back; but the battle must begin now.

2

Five Crises Looking for Heroes

In 1961 American president John F. Kennedy announced his Moonshot programme. He challenged the nation's engineers and aviators to put a man on the Moon by the end of that decade. They did just that when Neil Armstrong made his 'one giant leap for mankind' in July 1969. Sixty years on, The Earthshot Prize is challenging the world to undertake an even more momentous mission – nothing less than an Earthshot to save our home planet.

Founded by Prince William but led by a global team of environmental champions, The Earthshot Prize challenges the world's inventors, problem solvers, pioneers, concerned corporations, governments and citizens to come up with transformational solutions to five global crises. To find ways of protecting and restoring nature, of building a waste-free world, of fixing our climate, of cleaning our air and of reviving our oceans. Each year for the next ten years, five prizes will be awarded for the best solutions tackling each of the five Earthshots. Fifty prizes for fifty ground-breaking solutions in the most critical decade in history. Collectively, they will form the widest-ranging and most prestigious environmental prize ever created, with a total in prize money of £50 million.

Of course, The Earthshot Prize does not belong to one person. It is among the most ambitious environmental coalitions in history involving several hundred partner organisations and institutions, scores of experts, and dozens of leaders and influencers from around the world to make sure everyone notices and gets involved. And ultimately it is a challenge that must involve us all. For the aim is nothing less than to ignite a movement that delivers social and technological change on a global scale that transforms how we all live on Earth, creating a stable and thriving world for future generations.

This book works alongside The Earthshot Prize to explore how we got here, what our goals should be for the future, and how we might achieve them. But it begins with the problems. In order to find the right solutions, we first have to understand the crisis we face. The darkest hour, they say, is just before dawn. And there's every reason to believe we are now at our darkest hour. So while in the next few pages we must face the five crises head-on, we should neither despair nor look away. By recognising and understanding our current predicament we will provide ourselves with the guide to head more quickly towards the light.

Nature: Ready for the rebound?

On a dirt track deep in the rainforest state of Mato Grosso in Brazil, forest researcher Michael Coe gets out of his

SUV and spreads his arms. This, he says, is the front line. Behind him, on one side of the track, is cool, moist Amazon rainforest. It stretches for hundreds of kilometres through an almost intact indigenous forest reserve, where 6,000 Xingu people have lived for generations. The Xingu are famous for two things: in recent times, for protecting their forests in the face of advancing cattle ranchers and agribusiness barons; and a century ago for hosting (and perhaps killing) the eccentric British explorer Colonel Percy Fawcett on his doomed journey through the jungle to find the fabled lost city of Z.

But on the other side of the track, Coe gestures towards a very different world. In front of the tall New Englander lies hot, bare ground being prepared to grow soy beans on a farm the size of fourteen Manhattans. It is owned by the world's largest soy company, who ship their product across the world for animal feed. Probably your weekend roast chicken ate some, as it was fattened up for slaughter.

For the past decade Coe has been researching what happens at the boundary between these two worlds: where the rainforest meets agribusiness; where the most biodiverse ecosystem on Earth, with tens of thousands of plant, animal and insect species, is being ripped up to make way for a single species; and where prowling jaguars give way to combine harvesters. Here and now is the tipping point, he says. 'What happens here will seal the fate of the Amazon.'

Coe works with local researcher Divino Silverio, the son of a poor sharecropper who volunteered to assist at the research station Coe established on the farm, and now publishes his own papers. Their joint studies show how the climate abruptly changes at the boundary between farm and forest. How the farmland is several degrees hotter, and has a dry season that lasts several weeks longer. This is because the Amazon rainforest, the world's largest, makes its own climate. Every one of its hundreds of billions of trees sweats 500 litres of water a day, moistening the winds that blow above. That moisture makes clouds and rain that keep trees downwind watered and cool. Soy fields don't do that. So where the forest halts, the climate switches from warm, succulent and steamy wet to hotter and much, much drier. From a landscape where fires are rare to one constantly at risk of going up in flames. No wonder the farm has a hangar full of fire engines, ready to spring into action.

On Coe's front line, with a starkness and abruptness probably unparalleled anywhere on the planet, we can see standing side by side the two worlds we could make in the coming decades. Do we want the forests and their ability to moderate temperatures, make rain and maximise bio-diversity? Or do we want them removed for rows upon rows of soy plants to feed our livestock?

• • •

In 2020 we gained a new perspective on that dilemma. Covid-19 may have joined a long list of diseases that have spread from animals to humans in recent decades. The coronavirus paralysed human societies across the world, killing millions. The full repercussions are far from clear as this book goes to press. Will the virus be suppressed, succumb to vaccination programmes – or keep mutating and terrorising our ways of living? But while considering those questions, we should also ask where it came from – and why.

Many disease scientists have highlighted the role of deforestation in outbreaks of infectious diseases that reach us from animals. Serge Morand of the French National Centre for Scientific Research has found a strong link between deforestation in tropical countries and epidemics of both malaria, which kills around half a million people every year, and the highly contagious and usually deadly Ebola virus. 'We don't yet know the precise ecological mechanisms at play,' he says. But the suspicion is that replacing tropical rainforests with palm oil or soy plantations disrupts the complex interplay between species within ecosystems, pushing diseases to find new hosts.

Healthy ecosystems harbour many diseases that circulate within and between species. But when those ecosystems are damaged, this natural system falters. The barriers protecting us from diseases jumping between species break down – and outbreaks among humans become more likely. Luckily, most of the millions of viruses out there do not infect humans. The problem, as we have discovered, is that it can take only one outbreak to create havoc for humanity. And with viruses able to travel round the world in less than a day, we have little or no time to react.

It is too early to know for sure if Covid-19 spread to humans in this way. There is no obvious link between deforestation in China and the bats thought to have brought the virus to a food market in Wuhan. But we do know that twenty years ago forest fires set by Malaysians clearing land to grow palm oil drove hungry fruit bats out of the forest and on to farms, where they ate farmers' fruit and dropped faeces into pig pens. The faeces contained a virus similar to Covid, known as the Nipah virus. The pigs caught it, and so did the people looking after the pigs. The virus spread from bats to pigs to humans. More than a hundred people died, and more than a million pigs were slaughtered, before the outbreak was brought under control.

Fruit bats that ran out of forest habitat were also the likely cause of the 2014 Ebola outbreak in West Africa that killed at least 11,000 people. It began in the village of Meliandou in southern Guinea, where the dense forest had been largely replaced by coffee and cocoa farms. Investigators found that the first human to catch it was one of a group of children in the village who played in a huge hollow tree stump where a colony of bats also lived. The stump had been left behind when the forest was cleared by the villagers. It can be that simple.

• • •

Planet Earth is a 'living planet'. Its surface, whether green or blue, is dominated by living things. They run the place – or did. Over billions of years they established life-support systems that maintain the chemistry of the atmosphere, ocean waters and much else in states entirely different from how they would be without nature. Our air is breathable because living organisms keep it so, topping up the oxygen and cleansing it of pollutants. Our water is drinkable because living organisms constantly purify it. Those same organisms make soil, recycle nutrients to create new plants from old, and control the climate. The forests, grasslands, marine ecosystems and wetlands are the lungs and livers, hearts and kidneys of our world. Without this balance of nature, our planet would be uninhabitable for humanity.

Nature is resilient and has seen off past disasters such as asteroid hits. But we are at real risk of pushing nature beyond its powers to maintain our own stable, habitable world. Many species that shared the planet with us have disappeared, and huge numbers have been moved across the planet, disrupting ecosystems further – and perhaps releasing new diseases. Extinctions are part of nature, of course, but in the past half-century these disappearances have been at a hundred times the normal rate. Viewed from a distance, our impact on nature over the past few hundred years is comparable to that caused by the asteroid that saw off the dinosaurs.

Fewer than 3 per cent of natural ecosystems on land are reckoned to be truly untouched by human influence – a snatch of Congo rainforest, a piece of remote Patagonia in Chile, a stretch of the Siberian boreal forests, and not a lot more. We have bulldozed and burned, poisoned and ploughed half our forests for cattle ranches and palm-oil plantations, corn fields and rice paddy. We have drained most of our wetlands and blocked most of our rivers. Almost half our grasslands are ploughed up, turned to agriculture or built on. Many of the rest are fenced or bounded by highways, blocking migration routes and fragmenting the hunting territories of predators such as bears and lions.

Probably the big herds of migrating animals have suffered most. They need to move across the land, sometimes for hundreds of kilometres, to find food and water in changing seasons. But now, as they are hemmed in, five of the twenty-four known past mass animal migrations simply no longer happen. Others – from the wildebeest in East Africa to several of the caribou herds of North America and the saiga antelopes that once roamed the grasslands of Central Asia in their millions – are in deep trouble.

Nobody knows how many plant species have disappeared as the planet's ecosystems have been converted for farming. But we do know that among them are many of the wild relatives of the plants we rely on to feed us. And these plants matter. Three-quarters of the world's food comes from just twelve plant and five animal species. Plant breeders say their survival depends on being able to introduce genes from wild relatives to protect them against diseases and pests, or to help them grow in warmer or drier climates. That ability is being lost as the varieties die out.

In our concern for forests and the great grasslands, we too easily also forget the world's wet places. They, too, matter to nature. Yet, almost without noticing, we have dammed three-quarters of the world's large rivers, turning hundreds of raging thoroughfares teeming with fish into little more than water pipes, flowing or not flowing according to our demands for tap water, to irrigate crops or generate hydroelectricity. Every day, we take 10 billion tonnes of water from the world's rivers. That is more than a tonne for everyone on the planet. As a result, many of the world's major rivers no longer reach the sea for much of the year. The Colorado and Rio Grande in the United States, the Indus in Pakistan, the Nile in Egypt, the Yellow River in China, and many more, expire short of the shore. Most others have lost their seasonal flood surges that were critical to much of their wildlife.

Many other rivers are now little more than pipes for our sewage and industrial waste. Let's take one example. Until a few decades ago the River Citarum, the longest and largest river in west Java, Indonesia, drained an area of tropical rainforest. It was clean and full of fish. Today, it is known as the world's dirtiest river, awash with filth discharged from hundreds of textile mills and dyeing works along its banks. Its waters are laced with lead, mercury and arsenic. A ten-year plan by the government to clean up the Citarum has yet to make much difference. The water still changes colour by the hour, as different factories pour in different dye waste. Farmers can no longer take its water to grow rice paddy. The few fish that remain are unfit to eat. Fishing families that once harvested its waters now hold their noses and forage for floating garbage to sell.

Before they were taken over by engineers, many of the world's rivers fed their water into extensive floodplains and wetlands, which were the lifeblood of past great civilisations. Among them are the Mesopotamian marshes in Iraq, the probable birthplace of the Garden of Eden story. But in the past three centuries we have bypassed or drained 85 per cent of the world's marshes, bogs, peatlands, lagoons and lakes, desiccating huge areas of once lush floodplain. Most of the world's lakes and inland seas are dying.

Probably our most dramatic single intervention was to divert the flow of the great rivers that once fed the Aral Sea in Central Asia. Instead of replenishing this inland sea, they irrigated the world's largest expanse of cotton fields. Most likely you are today wearing some item of clothing made from this cotton. Until half a century ago, the Aral Sea was the size of Belgium. Today, if you look out from the beaches of its former holiday resorts and across its former fishing harbours, there is no water – only a new and largely unexplored desert, extending for more than 100 kilometres to a small, saline and fishless sump evaporating in the desert sun.

The United Nations (UN) has called the emptying of the Aral Sea 'the greatest environmental disaster of the twentieth century'. But it is part of a wider story of our reckless replumbing of the planet's water courses that has caused population sizes of freshwater wildlife to slump by over 80 per cent – a greater loss than for any other ecosystem.

Below ground, things are also bad. Soils are the planet's underappreciated ecosystems. A quarter of all animals live in the soil – from nematodes to moles. Soils take hundreds of years to form, as bacteria, fungi and other living organisms combine with wind and rain to break down hard rock and create a nutritious medium where plants can grow. Soils hold more carbon than all the forests and the atmosphere combined. But two-thirds of them are being degraded by erosion and destructive farming methods. Nature does its best to make good the loss, but we are losing topsoil ten times faster than living things can renew it. Every year, there is 30 million tonnes less topsoil. At that rate, we may only have a century or so of topsoil left.

And beneath the soil, nature's vast underground stores of water in porous rocks are being emptied. As we drain wetlands and diminish soils, nature is losing the ability to replenish these underground reserves by percolating rain-water downwards. At the same time, we are pumping up this water to fill our taps and irrigate our crops.

Once we drew water with buckets lowered down hand-dug wells; today, modern drills and petrol-driven pumps can tap the deepest water twenty-four hours a day. India empties its rivers for much of the year to irrigate crops and is now pumping ever more underground water. Millions would go hungry without it. But the pumping cannot go on. Water tables are falling by metres every year. Experts calculate that two-thirds of food production in some areas could soon be lost as the water runs out.

$$\cdot \quad \cdot \quad \cdot$$

This is a sorry, depressing and forbidding story. Under our recent stewardship, humanity has rarely valued nature and has all but ignored the services it provides to sustain our world. But if we look harder, there are some good news stories. Stories that can begin to shine a light at a future where we live in greater harmony with the natural world. For we are not on a one-way trip to disaster. Some human tipping points may have been passed, and we have started to reduce our destructive impact on the planet. There is now some modest recovery going on.

We may have reached 'peak farmland', for instance. By the year 2000, 46 per cent of the world's grasslands had been converted to agriculture, through being either ploughed or intensively grazed. But since then there has been a slight recovery, as yield growth on existing farmland exceeds population growth. Farmland has been abandoned, especially in Europe and North America, and is being recolonised by nature. As a result, while forest ecosystems continue to disappear in the tropics, often to supply Western nations, forest cover worldwide has stopped declining, with recoveries through planting and natural regeneration in temperate lands now exceeding deforestation in the tropics – in tree coverage though not in replacement of the tropics' species-rich biodiversity. The world has so far added an area the size of France to its tree tally this century.

Increasingly, parts of the world are being set aside for conservation and ecological restoration. In China and Russia, the United States and elsewhere there are extensive projects to restore drained wetlands. Germany is rewilding large heaths such as Lüneburg and Königsbrücker, which were once used for military training. The Dutch, who created much of their country by draining boggy land and 'reclaiming' the ocean, have stepped back from one of the big polders they created in the twentieth century, allowing nature to re-establish itself.

Wolves are returning to western Europe from the east for the first time in a century, creeping down railway tracks and sauntering through abandoned farmland.

Numbers of lynx, brown bears, wolverines, beavers and ibex are all increasing on a densely populated part of our planet.

As we will see in more detail later in this book, we often only have to step back for nature to recover. It won't always return to what it once was. Too much in our world has changed for that. We cannot protect every species or bring back what has been lost. But we can nurture nature's continuing persistence, and its ability to evolve and adapt to new circumstances. The question now is whether we will give it that chance, before the dynamism is exhausted. Will we retire our ploughs from the great grasslands, stay the chainsaw from forests, tear down the fences that block migrating beasts, end the pollution that is taking temperatures beyond anything nature has known for millions of years, dynamite dams that block our rivers, and remove the pumps that desiccate our wetlands? We can. And, as we shall see, we could live good lives while doing it. But will we?

Oceans: On the beach and over the horizon

Sailors off Vancouver Island, on the Pacific coast of Canada, have grown rather attached to three local pods of orcas that have swum in their waters for more than a decade, hunting the chinook salmon. They know the pod members by name, follow their health closely, and take tourists out to share their enthusiasm. So it made headline news in 2018 when Tahlequah, a twenty-year-old mother in the J-pod, was spotted tenderly carrying a dead calf. She balanced it on the top of her head for seventeen days.

For the world, it was a heart-breaking story of one animal's grief. But for researchers who follow the pods most closely, it was part of an increasingly disturbing pattern. Miscarriages and the deaths of newborn calves are now widespread among orcas. More than two-thirds of pregnancies in the Vancouver pods end in miscarriage, and almost half the calves that are successfully born die in their first year. These majestic marine mammals – technically dolphins, but also known as killer whales – can live as long as humans and form family groups much like ours. But they are falling prey to an insidious killer. Naturalist Talia Goodyear blames a group of industrial chemicals known as persistent organic pollutants, which stick around in the environment, getting absorbed by marine life such as plankton, and accumulating in often lethal quantities in animals at the top of the food chain.

The main culprits are polychlorinated biphenyls (PCBs). They were used in electrical components and added to plastics until being banned in Canada and many other countries in the 1980s. But most of the million tonnes of PCBs produced then have never been destroyed. They continue to leak into the environment, lingering long in the oceans – and poisoning orca foetuses. We rightly rage about marine animals snared by plastic bags and discarded fishing equipment. But often it is these hidden chemicals that are more lethal.

Naturalists such as Talia talk of an upcoming orca apocalypse. A recent study of the corpses of more than 300 dead orcas found some of the highest levels of PCBs in the fat-rich milk that mothers feed to their suckling calves. It concluded that this blight on reproduction could doom half the world's orca population by mid-century. Levels of poisoning are currently lower among orcas in the Arctic Ocean. But many persistent organic pollutants are known to be gradually accumulating in the cold waters there. So even they may not be safe.

· · ·

On the beach outside Monrovia, the capital of Liberia, one of the world's poorest countries, fishers gather. Like tens of thousands of others along the coast of West Africa, as far as Senegal and Mauritania, they are readying their traditional small fishing canoes to head out to some of the world's richest fishing grounds. In the old days this was easy work. They could fill their boats in a few hours and head home to sell their catch at beach fish markets that fed much of the country. But now the stocks of grouper, bream, tuna, sea bass and hake are disappearing. It can take days at sea before the fishers are ready to turn for home. And the entire journey is dangerous, because offshore, constantly cruising the fishing grounds, are giant trawlers. Their captains don't back off if they see canoes. Many local fishers have died after being mown down by the trawlers that are taking their fish.

The trawlers come from everywhere. As other fishing grounds decline, more and more make the journey to West Africa, where cold nutrient-rich water surfacing from the ocean depths can support massive shoals of fish in demand across the world. This time, in June 2020, there are five super-trawlers from China lurking offshore, catching Liberia's fish. Each measures some 46 metres from bow to stern and weighs 460 tonnes. They set nets tens of kilometres long. Each ship is capable of catching more than 2,000 tonnes of fish a year, depriving local beach fishers who reckon an annual take of half a tonne is good business. So one foreign trawler can take the fish from 4,000 local people. Overall, the trawlers threaten an industry on which 37,000 people depend, and which once fed much of Liberia's population of 5 million people.

In the recent past cash-strapped African governments have sold fishing licences to the foreign trawler owners. They feared that the boats would come anyway, with or without licences. But Africans are starting to get wise to the plunder, and the folly of selling their precious resource so cheaply. Reportedly, the five Chinese vessels off Liberia had already been forced to leave waters around Mozambique in the Indian Ocean, before heading west and being denied licences from Ghana. And in Liberia, too, they for the first time met opposition in government offices as well as on the Monrovian beach. The vessels began fishing and belatedly asked for a licence. But in September 2020 the director of the Liberian fisheries authority, Emma Glassco, turned down their licence request and told them to leave. Her refusal came just months after the government in Senegal had rejected licence applications from fifty-two foreign trawlers.

Would the Liberian ban hold? Liberia has been on a path

to democracy in recent years, but two decades ago it fought a bitter civil war that was largely funded by the proceeds of ransacking its other great natural resource, its forests. The rule of law in Liberia is frail. On the beach, they hoped Glassco's ruling would be a landmark for the country. A landmark that would allow them back into their waters without fearing for their lives – and with fish in the sea. But some months later, checking on how things had progressed, we found that one of the Chinese super-trawlers banned the previous summer was moored in Monrovia harbour.

• • •

If we want to rescue the oceans, we have to do three things. The first two are hard: prevent their poisoning, and stop their over-exploitation. Once dissolved in seawater, poisons know no boundaries. And giant fishing vessels can come and go before anyone is any the wiser. But the third is easier, and requires governments to look not beyond the horizon, but closer to home. To their own coastal eco-systems, which protect the nursery grounds where marine life breeds and grows.

Fed by silt and nutrients coming off the land, coral and mangroves, salt marshes, mudflats and kelp forests nurture the ocean fish on which a billion people depend. The fish-eries off West Africa, for instance, are fed by spawning grounds in coastal lagoons, mangroves and muddy bays such as the Banc d'Arguin in Mauritania. Protecting them will go a long way towards restoring the oceans.

Of all the coastal ecosystems, coral reefs are the most

revered. These vast living structures are composed of trillions of soft-bodied coral animals whose external skeletons fuse together to make the reef. Inside the shelter of the coral, constantly feeding them, live multicoloured algae. The symbiotic relationship between coral and algae creates a reef world in which all manner of sea life can thrive. Though they cover only 0.1 per cent of the world's oceans, coral reefs are home to about a quarter of ocean species, and probably play a role in the life cycle of half of them.

Coral reefs are the largest and longest-living creations of nature. The Great Barrier Reef off the east coast of Australia is 2,000 kilometres long and 5 million years old. But even it is a relative newcomer. The coral at Enewetak in the Marshall Islands of the Pacific Ocean is 60 million years old, and extends for more than a kilometre down to the ocean bed.

We are doing huge damage to many of these great structures: scraping them with ships' anchors; smothering them in pollution; dynamiting or poisoning or flailing them with nets as we hunt their fish. Still, they are in many ways, tough. The coral at Enewetak and nearby Bikini Atoll regrew after nuclear bomb tests were conducted in their midst by the United States in the 1950s. Some 80 per cent of coral species returned. We saw earlier in Mexico's Gulf of Cortez how the reefs there have regenerated and their inhabitants have come back after a marine protected area was created.

But many coral reefs seem extremely sensitive to changing ocean temperatures. Sudden warming – typically a combination of long-term climate change and short-term local warming – causes the algae to flee the coral. Naked of the algae and starved of nutrients, the iridescent colours of the coral turn to a deathly white. Scientists call this coral bleaching. The Great Barrier Reef suffered a mass bleaching in 2016 and again in 2017 and 2020.

This need not be the end for the reef. If the warming is short-lived, the algae will return and the coral can recover. Some coral species, such as those in the Red Sea, are more resilient than others to sudden temperature change. The larvae of others can migrate to cooler waters, floating on ocean currents for several months until they find some new coastline that is more hospitable. They have been recorded moving north up the Florida coast, and south from Australia's Great Barrier Reef. Warnings that all tropical coral reefs could be dead by mid-century may be wrong. But there can be little doubt that many of these ancient stable reef structures, and the life within them, are in deep trouble, if warming continues.

Not far behind coral reefs in their importance for marine ecosystems and tropical fisheries are mangroves. These stubby salt-tolerant tropical trees grow along thousands of kilometres of intertidal mud in more than a hundred countries. They harbour sponges and worms, shrimps and sharks, crocodiles and crabs. Possibly half of all tropical fish are born among their roots.

Away from the tropics, mangroves give way to seagrasses that form lush offshore meadows around all continents except Antarctica. Even more widespread, and a bit further offshore, are the 'forests' of kelp, a giant seaweed that is capable of growing half a metre a day and reaching up to

45 metres in height, spreading their canopies of foliage just below the water's surface. Inshore of seagrasses, many low-lying coastlines are lined with salt marshes that are colonised by all manner of salt-tolerant bog-loving herbs and grasses.

For millennia, coastal communities at all latitudes harvested these different coastal ecosystems without causing them much long-term damage. But various combinations of outright destruction by farmers, disruption of ecosystems by pollution or fishing, and climate change are placing many under serious threat in the twenty-first century.

Sometimes their disappearance has been sudden. The kelp forests that luxuriated off the shore of northern California were largely intact as recently as 2014, but six years later 95 per cent were gone. They had faced a perfect storm of threats, including unusual ocean warming and an invasion of sea urchins (which graze on kelp) when disease ripped through their main predators, starfish.

Sometimes they are eaten away gradually. During the last quarter of the twentieth century British sheep farmers and engineers drained a fifth of the salt marshes that once extended either side of the Thames estuary. Far away in Indonesia, coastal communities dug out 40 per cent of their once extensive mangroves, to make room for ponds to farm prawns and milkfish.

All told, we have lost roughly half of these rich coastal ecosystems. And whether slow or quick, the effects can be devastating. After Indonesian villagers in north Java got rid of their mangroves, the sea invaded the low-lying land behind. It spread for several kilometres through row after row of ponds and rice fields, drowning some villages and leaving others connected to the mainland by narrow causeways that villagers raised up each year to prevent them being washed away.

Their loss also haemorrhages more carbon to the atmosphere than deforestation on land. Hectare for hectare, mangroves, seagrasses and salt marshes hold more carbon than the densest jungle. But most devastating, perhaps, has been the impact of their loss on ocean fish stocks. The supply of fish from coastal ecosystems into the deep ocean has faltered just as fishing on the high seas has peaked. The result has been a law of diminishing returns for fishers.

More and more vessels cast bigger and bigger nets ever deeper into the oceans. Gill nets dragged by two or more fishing vessels can be more than 30 kilometres long. These 'walls of death' were banned by the UN three decades ago, but Greenpeace has exposed how they are still in use, harvesting everything in their paths in the Indian Ocean and elsewhere. Baited longlines are little better. They can be tens of kilometres long, studded with hooks the whole way. Unwanted species including shark, turtles and seabirds make up a fifth or more of what is caught in such nets and on such lines. Yet, despite all the effort, these industrial fishers are catching fewer fish than they did twenty years ago. Because the fish are simply not there any more.

The Pacific bluefin tuna – which lives for up to fifty years, can grow as big as a horse and fetches up to a million dollars in Tokyo's main fish market – has seen its numbers fall by 97 per cent. Two-thirds of sharks are gone too, their corpses often discarded once their fins have been cut off to make shark-fin soup. And through the twentieth century, in one of its most shameful slaughters, humanity has emptied the oceans of most of their greatest creatures, the thirteen species of great whales.

Whales, the planet's biggest animals, are vital ocean gardeners and recyclers. By diving to the depths to feed, and then surfacing and pooping, they redistribute nutrients in the oceans, maintaining life at the surface. When they die, their bodies sink to the ocean floor – a process called 'whale fall' – taking to the depths the huge stores of carbon their bodies are made up of – a very literal 'carbon sink' – and providing nutrients for new life on the seabed. 'Whale fall' are oases of life and activity in an otherwise bleak seascape. A dead whale can provide food for eighty years or more.

Or that is how things used to be. In their heyday, perhaps 5 million humpback, minke and blue whales cruised the oceans. But their numbers collapsed in the twentieth century as we hunted most of the great whales almost to extinction – to provide whalebone for corsets, and blubber to make oil for candles, street lights, lipstick and, especially towards the end, margarine. In the 1950s and 1960s as many as 50,000 great whales a year were being harpooned and dragged by cranes into the holds of giant whaling ships. The biggest of them all, the blue whale (think of a creature the size of thirty elephants), probably suffered more than any, till a halt was finally called to their slaughter in 1966, before all great whales won protection in 1986.

• • •

The oceans cover two-thirds of the Earth, and in places are more than 10 kilometres deep. They are where life on Earth began, and are by far the largest living spaces on the planet. They also perform myriad services for the rest of the biosphere. Water evaporating from their surfaces in the sun makes our rain. Ocean plants known as phytoplankton have produced most of the oxygen in the atmosphere. The oceans and the life within them are also vital controls on the planet's temperature, absorbing four times more carbon dioxide than the Amazon rainforest.

Global warming would be much worse without the oceans to dampen the impact. But ocean's capacity to take up carbon dioxide has a downside. Dissolved in water, the gas becomes carbonic acid. The uptake is now so great that our oceans are gradually becoming acid. Many ocean creatures have shells and skeletons made of calcium carbonate, which is dissolved by the acid. Coral reefs may be eaten away. Effects so far are minimal, but marine biologists predict major damage to marine ecosystems by mid-century if carbon dioxide levels in the atmosphere continue to rise, pushing ocean uptake ever higher.

Almost nowhere in the oceans is untouched by human influence. A plastic bag was recently found at the bottom of the Mariana Trench, the deepest spot in the world's oceans. The threats extend far beyond fishing and toxic chemicals and plastics, to less well-known but perhaps more insidious intrusions. These include the floods of nitrogen fertiliser creating dead zones in shallow coastal seas, which we discuss later; and underwater noise from ships' propellers, oil drilling and offshore wind turbines that drown out the communications between whales and many other creatures, leaving them disoriented, stressed and sometimes starved. Noise may be one reason for whale strandings on our coasts.

In places, complex food webs dominated by big animals are being turned into ecosystems where jellyfish and toxic algae rule. Pessimists fear the worst. 'We can only guess at the kinds of organisms that will benefit from this mayhem,' said Jeremy Jackson of the Scripps Institution of Oceanography a few years ago. 'Microbes will reign supreme,' he predicted.

Maybe. But there is good news, too. There have been an increasing number of international agreements aimed at limiting the plunder of everything from South Pacific tuna to North Atlantic cod, and African sea bass to the Patagonia toothfish that swim the waters around Antarctica. Local no-fishing zones are being complemented by giant protected areas in more remote parts of the world's oceans – around far-flung islands of the South Atlantic, North Pacific, Indian Ocean and in the Southern Ocean around Antarctica. At the time of writing, the UN is negotiating a high seas treaty aimed at bringing more of the oceans under inter-national law.

Not all these initiatives work well yet. Policing the high seas is hard. Thousands of sharks have been caught illegally around the Chagos Islands, in the Indian Ocean, since the British government established a marine protected area there in 2010. But there is progress. 'The nadir of over-exploitation came about in the mid 1990s,' says marine biologist Callum Roberts, now of the University of Exeter.

Some wonder what took the world so long. After all, nations have an interest in maintaining fish stocks off their shores. So do fishing companies that have to send their ships ever further and on ever longer voyages to bring in their catches. Still, better late than never. And it is not too late, according to a report by Roberts and a team of elite experts in 2021. Better pollution controls, 'common-sense' management of fish stocks and some hefty protected areas, including around coral reefs and other coastal fish nurseries, could restock the oceans by 2050, they say. As we saw in the Mexican fishing village of Cabo Pulmo, ocean ecosystems can recover quickly – faster than many terrestrial ecosystems. And recovery can be catching, as restored coral reefs and mangroves act as nurseries that supply fish far and wide.

But for now, at least, the high point of marine protection remains what happened almost four decades ago. In 1986 the International Whaling Commission, a club of whaling nations, called a belated halt to their activities. Facing mounting public pressure to 'save the whales', the commission imposed a global moratorium on commercial whaling. In truth, by then whale numbers were so low that whaling was barely profitable any more, and the pressure to call a halt was one of the earliest and strongest messages of the growing environment movement. Still, thirty-five years on, the moratorium remains in place.

Recovery takes time. Whales live for many years and breed only slowly. Even so, after being down to fewer than

3,000, blue whales now number between 10,000 and 25,000. That is between a twelfth and a fifth of their numbers a century ago. Humpback whales are doing even better. Since 1986 their populations have grown tenfold to at least 120,000. Once again, they are embarking on their great annual migrations, from the rich polar waters where they feed, to the tropical waters where they breed. Another tenfold increase and they will be back to their pre-harpoon numbers. If we can bring back these denizens of the deep, then we can surely bring back any creature.

Air: Choking on progress

Tumursuuri Chuluunbat is a country man from the wind-swept steppes of Mongolia, a landlocked nation, sandwiched between Russia and China, where almost everyone once lived in remote rural communities herding cattle and sheep. But, with climate change and giant mining companies turning their lands to dust, he is one of some 2 million people, half the country's population, who have moved to the country's fast-growing capital, Ulaanbaatar.

The migrants huddle together for warmth in tented ghettoes in the coldest capital city on Earth, where night temperatures can be as low as minus 40 degrees Celsius. Their only source of warmth is the product that Tumursuuri sells from a small truck: coal. And it is killing them.

The effect of coal burning on the city's atmosphere is terrible. Ulaanbaatar is also one of the smoggiest cities on Earth. The coal fumes create black clouds that keep ever more pollution trapped in streets and whistling through the felt tents, known here as *gers,* where two-thirds of the city's population live. 'Even in our house, the smoke fills the rooms,' says Tumursuuri.

The story of Ulaanbaatar's smogs is an international scandal. Government sensors in January 2018 recorded levels of the finest, most dangerous particulates at 130 times World Health Organization (WHO) safe levels. After an outcry, the sale of raw coal was banned and replaced by coal briquettes, which is what Tumursuuri now sells. But the quality of the briquettes is very bad, he says. 'There is still a lot of coal dust.' On bad mornings, visibility in the city can be as little as three metres. And the odour is terrible, he says. 'It smells like sulphur. It give us headaches and sore throats, and makes our children vomit.' He spends his nights at home fearing for the safety of his five young children.

Tumursuuri came to Ulaanbaatar because it was the only way to get his children an education. 'But we are sacrificing our health,' he says. In his neighbourhood, respiratory diseases such as bronchitis, pneumonia and asthma are rife. Pneumonia is the second biggest cause of death among under-fives here, according to UNICEF. He dreams of going back to the countryside, where he could work again as a woodcarver. But till then, he prays for better air. 'We Mongolians worship the eternal blue sky,' he says. 'Every morning we pray to the sky for the protection of our children by offering milk and tea. Yet we are polluting it. I look forward to the day when we can see clear skies again.'

• • •

The atmosphere is our most precious resource. We must breathe it every minute of every day. And it depends on nature. Living organisms constantly renew its complex chemistry. They maintain oxygen levels sufficient for us to breathe and levels of carbon dioxide and other greenhouse gases sufficient to keep us warm, but not so warm we expire. The circulating air brings us rain for water, and the air aloft provides an ozone layer that shields us and all life on Earth from deadly solar radiation.

But the air we breathe is increasingly defiled. Our pollution sometimes overwhelms nature's powers to cleanse it. Traffic emissions, soot and smoke from coal burning and dirty industries, forest fires, tiny particles that fly off tyres, all pollute the atmosphere. They damage ecosystems, alter climate, fall into the ocean, and lodge in our lungs.

Billions of us breathe polluted air every day. The health effects are devastating. Air pollution contributes to an estimated 7 million early deaths annually round the world. That is 20,000 deaths every day – more even than Covid-19 at its 2021 peak. As countries such as Mongolia industrialise, burning dirty coal in millions of power stations, industrial grates and home fires, and filling the streets with fumes from millions of vehicles, the smogs worsen. Chinese scientists calculate that their citizens lose an average of three years of their life to choking air. And while Beijing, Shanghai and other major Chinese cities have somewhat cleaner air these days thanks to clean-up programmes, Indian cities are still becoming filthier. Of the world's thirty most smoggy cities, at least twenty are in India, with New Delhi often the worst.

Such dense smoky smogs no longer happen much in most rich countries, where coal burning is kept out of cities and power stations have high chimneys that spread their effluent on the winds. But the air still fills with the vehicle emissions that release toxic gases and tiny particles right where our children breathe it in. For a long time, most of us rather ignored this unpleasant truth. Then, in the UK at least, came Ella.

Ella Adoo-Kissi-Debrah, like millions of other children, suffered from asthma. The nine-year-old schoolchild took an inhaler with her every day to school in Catford in south London. In the middle of a smoggy week in February 2013, she died after a bad asthma attack. But for the determination of her mother Rosamund to find out the truth behind her death, she would have been a forgotten statistic. Instead, seven years later, she became the first child in Britain where a coroner recorded on the death certificate that car exhausts had made a 'material contribution' to her death.

Philip Barlow noted that Ella lived close to one of London's busiest roads, the South Circular, and that traffic emissions were 'the principal source of the air pollution that caused her worsening illness over three years'. He sent a formal request to the government calling on ministers to report back to him on how they were implementing WHO guidelines on air pollution, which were regularly exceeded in Catford, as elsewhere in London.

This was not just about Ella or her classmates, he said. There was 'no dispute . . . that atmospheric air pollution is the cause of many thousand premature deaths every year in the UK'. Nobody could say exactly how many die like

Ella each year. But one scientific witness at the inquest, Stephen Holgate of the Medical Research Council, put the annual death toll in Britain at a staggering 40,000. Air pollution in London is reckoned to have a role in one in five deaths, from asthma, bronchitis, heart attacks, strokes, cancer, dementia, diabetes and more. That is a lot of causes of death to be missing from decades of death certificates.

Ella's fellow pupil at school before her death, Anjali Raman-Middleton, remembers that in her school 'lots of people had asthma. Having a pump felt as normal as wearing glasses. It was almost part of the culture. I didn't realise that asthma was something that could kill you. We never thought twice about the air we were breathing.' When Rosamund learned about the danger of air pollution and began to campaign for an inquest, they set up a campaign group, called Choked Up, to demand cleaner air. 'Ensuring every child, woman and man in the UK has clean air to breathe should be a government priority for the sake of our health, our economy and our climate,' she says.

By chance, Ella was Black. Or maybe not by chance. Research from the United States published in 2021 reported that nearly every source of the deadliest air pollution in the country disproportionately affects Americans of colour. Decisions made over the decades about where to build highways and factories mean that neighbourhoods with more people of colour are exposed to 10 to 20 per cent more soot and dangerous particulates than others. Anjali says London is the same.

We are learning more all the time about the cocktail of chemicals polluting our air. Most of us would give little thought to the black smoke from the chimneys of brick kilns. But there are an estimated 100,000 brick kilns worldwide. Countries like Bangladesh, India, China, Pakistan and Vietnam have tens of thousands of them. The world as a whole makes 1.5 trillion bricks a year from polluting kilns – 200 for each of us. But it has only recently become clear that brick kilns probably account for a fifth of all the soot particles in the atmosphere round the world.

And closer to home, there are other until recently unsuspected hazards. For instance, our city air is full of particles of rubber and plastic created as tyres disintegrate in contact with roads, especially when vehicles brake or turn sharply. All motorists know that their tyre treads degenerate until they become bald. Nobody ever thought to wonder where the bits that flew off went. But a study for the British government estimates that they make up around two-thirds of all the tiny particles reaching the air from road traffic, making them potentially as dangerous as diesel emissions. Hugo Richardson of The Tyre Collective, a group of engineers highlighting the issue, says that 'just in Europe, tyres produce over half a million tonnes of pollutants every year'. They may be the second biggest source of microplastic pollution in the environment. The bits that don't end up in our lungs may float in the air for days before falling on to fields, rivers or the ocean.

• • •

Urban areas may suffer most from air pollution. After all, they emit most of it. But nowhere is safe. While today's air pollution is less visible than that of previous decades even now the emissions from faraway cities get transported by air currents to remote communities. Looking to historical impacts can help us understand what is happening today.

To banish twentieth-century smogs, cities across Europe built a new generation of giant coal-fired power stations out of town, with tall chimneys to spread their pollution. The result, through the 1970s and 1980s, was that the sulphur in the smoke became incorporated into clouds and turned the continent's rain into acid – sulphuric acid. The acid killed forests across Germany and central Europe, and fish in thousands of lakes across Scandinavia. It was a huge environmental issue – as important then as climate change is today. Power companies were eventually forced to put filters on their chimneys to remove the sulphur pollution. But the acid rain scandal proved one thing – air currents don't respect borders. When we put pollutants into the air, they travel far and wide, within days. The actions of some of us will affect all of us, and, ultimately, all of life.

In the tiny Swedish township of Jokkmokk, deep inside the Arctic Circle, Jannie Staffansson tells how her indigenous Sami people, and the herds of reindeer that are central to their lives, have suffered from air pollution though remote from the rest of the world. First it was acid rain and soot from British and German industry, then the radioactive fallout of the Chernobyl nuclear disaster in Ukraine. Today, the air pollutants of the growing economies of Eastern Europe and Central Asia once again prove how connected we all are.

'Reindeer are everything to us,' Jannie says. 'They are our creation myth. The Earth was made from reindeer. Their eyes are the stars. Their fur makes the forest. Their blood makes the lakes. I go to pieces every time one of my reindeer dies.' What is scary about air pollution, she says, is how greatly you can be affected by the emissions of others far away.

Reindeer are 'canaries' – that is, they are harbingers of the invisible dangers that lie in the air. Their chief foodstuff is, aptly, the 'reindeer lichen' that covers the floor of boreal forests and the open plains of the tundra. Lichens are famous among environmental scientists as the best biological indicators of air quality. This is because lichen have no roots and acquire all their nutrients through their surface, so they are poisoned easily. Different lichens are so sensitive to particulates in the air, an expert can predict the quality of the air about them purely from the types of lichen found on gravestones, rocks and trees nearby. To a certain degree, anyone can do this. Explore your local area. If you can find bushy, leafy lichens, the air quality about you may be quite good. If the lichens are no more than crusts hugging the rocks and walls, it's a sign of poor air quality. If there are no lichens at all, it might be time to think about starting a campaign.

Jannie's reindeer start to run out of lichen whenever the winds blow in new pollution. 'We reindeer herders live close to nature. Those who live closest to nature are affected first, and maybe disappear first,' she says. Her words hold two stark truths. The first is that inequality is a key component of the problem of air pollution. It is often those least responsible for the pollution that are most affected. Globally, WHO has discovered that children, the elderly and the poor are the most exposed. In low-income countries, the regulation of energy generation, industry and domestic fuels is routinely inadequate. Those in richer nations all too often scorn nations with high levels of air pollution, but they have to recognise that many of the goods manufactured in the poorest and most polluted parts of the world are bound for wealthier markets. This 'offshoring' of air pollution ought to cause international outrage, yet it is routine.

The second stark truth is that nature suffers alongside us. If reindeer are canaries, habitats and species across the world must also be affected. Yet, while we are now beginning to measure how human health is affected by air pollution, we are a long way from understanding the many detrimental effects our dirt and gases must be having on the natural world.

As well as innovations within cities to tackle air pollution, we will clean our air by enacting global standards on manufacturing and industry. We have done this once before and the results were remarkable.

• • •

High above our heads in the stratosphere, ozone forms a protective shield that protects life below from cancer-causing ultraviolet radiation streaming towards us from the sun. This ozone layer has been under assault from a group of man-made chemicals that contain chlorine compounds. These ultra-stable compounds rise up through the air and once in the cold stratospheric air, they rapidly destroy the ozone.

The worst of these chemicals, chlorofluorocarbons (CFCs), were invented in the 1930s as a coolant for refrigerators. By the 1970s their concentrations in the upper air had reached levels high enough to burn a hole in the ozone shield over Antarctica. When researchers noticed the hole, in 1985, there was panic. Few had predicted that this runaway effect might happen and those that had were ignored. There were fears that a similar hole could open up above the Arctic, exposing hundreds of millions of people in nearby countries.

But the world acted. Two years later, the Montreal Protocol agreed a programme to ban the chemicals. Since then, the ozone layer has slowly started to heal. It will likely be late this century – a hundred years after the hole was discovered – before it is finally resealed. Still, at least the world showed it could act together in the face of an immediate environmental crisis. If we can fix the ozone layer, say atmospheric scientists, we can fix both air pollution and climate change too.

Climate: Back from the brink?

The Maldives is the lowest country in the world. An archipelago of 1,200 coral islands stretching across the Indian Ocean south of India, it is an earthly paradise. But it is in grave peril as climate change causes sea levels to rise and ocean storms to worsen. The inhabitants of the capital Male are probably safe for now, behind a three-metre sea wall. But for the rest of the country's 400,000 inhabitants on around 200 other islands, few of which reach more than a metre above current sea levels; their homes look temporary.

It is hard to be precise about how fast sea levels will rise in the coming decades. But as the oceans warm, their water will take up more space; and as the atmosphere warms, glaciers and ice sheets on land across the planet will thaw, adding further to the volume of water in the oceans. Sea levels are currently rising by more than 3 centimetres a decade. If we are lucky, that means half a metre this century, but some scientists predict that without early drastic action to halt the warming, it could easily be more than a metre. Some reefs, if still healthy, have so far kept pace with rising sea levels. But this may not continue. Even if they are not permanently engulfed, many low-lying islands and coastal regions face becoming uninhabitable as storms and high tides buffet communities, infiltrate water supplies with salt, and break sea walls. But even if we are lucky, it won't stop there. Sea levels will keep rising for centuries as ice sheets and oceans find a new 'normal'.

For the 4,000 people who lived on the tiny but densely populated island of Kandholhudhoo in the north of the Maldives, home has already gone. The 500-metre-long island was completely inundated by the tsunami that washed across the Indian Ocean in 2004. There was a warning and all but three of the island's inhabitants escaped. But its coastal defences were washed away, so it has never been safe to return. Kandholhudhoo is now a ghost island.

Its former inhabitants are exiles on nearby Dhuvaafaru. They want to return and rebuild their lives. The want to invite back the long-haul tourists who once dived amid the coral. Tourism is the main business on the Maldives these days. But the former inhabitants of Kandholhudhoo know that the emissions from the flights that would bring back the visitors would also contribute to the warmer world that is raising tides here. The emissions would mean more flooding of their endangered paradise. It is a horrible dilemma.

• • •

If danger for the people of the Maldives comes from the sea waters lapping at their toes, in the Peruvian town of Huaraz, the threat of climate change comes from above. The city of 120,000 people is high in the Andes Mountains, and is constantly at risk from a lake perched above it. The volume of water in Lake Palcacocha has grown thirtyfold, to more than 17 million cubic metres, as it fills with water from glaciers in Peru's Cordillera Blanca, among the fastest-melting glacier regions on Earth. American glaciologist Daene McKinney, who has studied the lake in detail, says a landslide or icefall into the lake could send a 30-metre-high wall of water spilling over its banks into the city below.

This is not just a theory. Something similar happened once before, in 1941, when the lake was much smaller than it is today. Even that event killed at least 1,800 people. In recent times, efforts to siphon off water from the lake have failed, and few in the city have confidence that they will get much notice if the worst happens. The only thing standing between them and death would be the alertness of a lonely watchman sitting in a hut by the lake, ready to phone down if anything happens.

Even if all went well, the people of Huaraz would have less than thirty minutes to escape. If a breach happened at night, they would likely all drown in their beds, says Saúl Luciano Lliuya, a farmer and guide in the Cordillera Blanca. He makes regular trips to Germany, where he has taken the country's second largest power company, RWE, to court, demanding that it pay part of the cost of making his city safe. He waves copies of a study showing that RWE is responsible for 0.47 per cent of the global emissions that are melting the glaciers above Huaraz. So he wants it to contribute that same percentage of the cost of safeguarding Huaraz. He sees the action as a test case for other actions against bigger companies in other countries that may make a bigger contribution to climate change. RWE contests that it has corporate liability and argues that ending its emissions would not reduce the threat to Huaraz.

There are hundreds of other remote lakes in the mountain regions of the world that are swelling as the glaciers above them warm and melt. Many of the most dangerous are in the Himalayas, where one gave way in early 2021. Luckily there was no city or town downstream. Even so, the burst killed around 170 people working at a hydroelectric dam in its path. The speed at which the world's 200,000 glaciers are melting has doubled in the first two decades of the twenty-first century. From Alaska to New Zealand, the Alps to the Andes, they lost 4 per cent of their volume in that time. Sometimes the water simply flows to the ocean, raising sea levels. But if that route is barred and a lake forms, the risks for communities downstream from a sudden breach of the lake's banks will be ever present.

Saúl's legal action in Germany is a test case for what could become a cascade of legal battles to protect other places from melting glaciers, rising tides, worsening storms, droughts and other catastrophic potential consequences of climate change. Though all agree it would be better to spend the money preventing climate change in the first place.

• • •

In Alaska, the world of Native Americans is changing, says one of their own youth campaigners Ruth Miller. Climate change is upsetting their relationship with the land and nature. 'The plants are changing as winter comes later; our caribou herds and whale pods are changing their migration patterns; new species foreign to these lands are moving north; and with the increased heat, wildfires are burning longer and hotter,' she says. 'For the people who live a traditional life – following the caribou herds and relying on salmon runs into our bays and rivers – supplies of food are no longer secure.'

It is not just food supplies at risk as their world warms.

Whole communities risk falling into the ocean, she says. Shields of thick sea ice once protected their coastal villages from fierce winter storms lashing at their shores. But now earlier melting is leaving homes exposed. The danger is compounded because thawing permafrost is leaving once rock-hard shorelines open to rapid erosion by the waves and tides. Faced with annihilation of their homes, the inhabitants of Newtok began moving to higher ground on nearby Nelson Island in 2019, and will have moved every home by 2023. Another village, Kivalina, decided to try to stay put, moving houses one by one on to a narrow spit of land close to the old village. But that spit is now itself disappearing into the Chukchi Sea.

Away from the coast, the thawing permafrost is changing the Alaskan landscape just as fundamentally. As ice turns to water, thousands of lakes and ponds form; meanwhile shallow-rooted trees fall over, roads buckle, and buildings sink and become unstable. There are global impacts too. The permafrost has long trapped methane gas from decaying vegetation. But as thawing gets under way, the methane is leaking into the air from pores in the ground and bubbling up through lakes. Methane is a greenhouse gas. Molecule for molecule, it is many times more potent at warming the atmosphere than carbon dioxide.

Katey Walter Anthony of the University of Alaska is documenting methane emissions bubbling up through the waters of a new lake, Big Trail Lake near Fairbanks, itself created by thawing permafrost. The methane is concentrated enough that she can light a flame as it bubbles out of the water.

These releases are 'a ticking time bomb', says Ruth. 'They will exacerbate every other kind of climate threat.' Methane released from permafrost has the potential to generate warming that will accelerate its release further – a runaway effect that could be impossible to halt. Some say the tipping point may already have been crossed. Jørgen Randers, a climate researcher at the Norwegian Business School, has produced modelling that suggests we now face 'self-sustained melting of the permafrost for hundreds of years', even if all further CO_2 emissions were stopped immediately. He got some flack from colleagues who criticised his modelling. But it shows we may be on the edge.

• • •

A stable climate has been essential to humanity's rise over the past 12,000 years, from a few million hunters and gatherers to today's global digitised society of approaching 8 billion people. A climate that was predictable from one year and decade to the next allowed us to plant crops with confidence that they would be watered by the rains. It gave us the confidence to build our homes and infrastructure on shorelines and beside rivers, knowing they would be safe from floods and storms. It is all we have known.

But all that is changing because we are tweaking the planet's thermostat. We are doing this by messing with the planet's carbon. We are digging and drilling carbon-based fuels such as coal and oil from the ground, and burning them to make our energy. In the process, we are taking carbon from its established place in rocks, and pouring ever more of it into the atmosphere. The concentration of the gas in the air is now approaching 50 per cent more than in pre-industrial times. Its 'greenhouse effect' has raised temperatures by an average of over one degree Celsius. The increase is greater on land than over the ocean, and twice as great in the Arctic, where melting ice and other feedbacks supercharge the warming. Temperatures are now higher than humanity has ever experienced, and without action one degree of warming will become three or four degrees by century's end, and more thereafter.

The warming is also altering the great flows of air in the atmosphere that create our day-to-day weather, destroying the predictability of rainfall and the seasons and creating unseasonal droughts and heatwaves, floods and storms. Climate change appears to intensify El Niños, the periodic reversals of ocean currents and winds across the tropical Pacific, which now bring drought and fire to the rainforests of South East Asia and floods to the deserts of the South American coast. The jet stream that drives weather systems across Europe is faltering. The Indian monsoon, on which a billion people depend to irrigate their crops, is becoming stronger but more erratic. German climate researcher Anders Levermann calls it 'a threat to the well-being of the Indian subcontinent'.

We have always had bad weather; local climate records have always been broken. But extreme weather events are becoming more frequent and more extreme. In the process, crops fail and ecosystems cannot cope. Rainforests succumb to wildfires and turn to grasslands. Some grasslands turn to deserts. Though it is also true that warmer times are encouraging trees and grasses to push into Arctic tundra and up mountainsides that were once too cold. So while there are potential gains as well as losses, the end to predictable climate may be the greatest peril of all.

Similarly, while some places will become warmer and more habitable for humans, many more are set to become uninhabitable, says Tim Lenton of the University of Exeter. 'A substantial part of humanity will be exposed to mean annual temperatures warmer than nearly anywhere today,' he says. 'Over the coming fifty years, between 1 and 3 billion people are projected to be left outside the climate conditions that have served humanity well over the past 6,000 years.' Heat and humidity will within a few decades make life unbearable, and potentially lethal without air conditioning, in much of the Gulf region, India, parts of Africa around the Sahara desert, including its most populated country, Nigeria, and even Australia.

As oceans warm, we are seeing ice melt, sea levels rise and land disappearing beneath the waves. Even with giant sea walls to protect cities and other densely populated coastlines, up to 70 million people face being forced out of their homes by rising tides before the century ends, with sea level rise set to double if the Paris Agreement to limit warming to 1.5 degrees Celsius is not met. Inundated islands, such as in the Maldives, may make the headlines. But much greater numbers of people would be flooded from their homes on the great river deltas of South and East Asia, such as the Ganges in India, the Brahmaputra in Bangladesh and the Mekong in Vietnam.

Most alarmingly of all, continued warming will unleash

forces that cannot be reversed. Critical tipping points may soon be crossed. One is the release of methane from thawing permafrost. What is happening in Alaska is also happening across northern Canada, Scandinavia and wide areas of Siberia. Another potential tipping point is the sudden release of ice on land, creating unstoppable rises in sea levels.

An unmanned research submarine sent in 2021 to one of the most remote places on the planet – the Thwaites Glacier in West Antarctica – found the ocean waters washing round its outer edge are warmer than previously thought. These waters are detaching the ice from the seabed. The kilometre-deep glacier acts as an ice plug holding huge areas of the West Antarctica Ice Sheet on land. If it became unhooked from the seabed beneath it, it would float away and likely take a large part of the ice sheet with it. This could eventually raise sea levels world-wide by more than 3 metres, the researchers warned. It is why they have called the Thwaites Glacier the 'Doomsday Glacier'. Once, we used the word 'glacial' to describe very slow processes. That seems about to change.

Other potential tipping points in climate change include a terminal decline of the Amazon rainforest. Brazilian researcher Luciana Gatti has been carrying out research flights over the Amazon for the past decade and declared in 2020 that 'we have hit a tipping point'. Part of the world's largest rainforest, which has for thousands of years been soaking up carbon dioxide from the air, is now releasing that accumulated carbon. 'Each year it gets worse,' she says. As we saw on the soy farm in Mato Grosso, once the

Amazon rainforest starts to disappear at the edges, the bare ground exposed results in higher temperatures and reduced rainfall, killing more forest and triggering more fires. Indeed, recent research indicates that the increase in deforestation between 2012 and 2019 has caused a 39 per cent increase in the number of fires across the Amazon. Now the entire Amazon system could be approaching a tipping point. Beyond it, even without further deforestation by ranchers and farmers, the forces unleashed by climate change will in turn release yet more carbon into the air and turn the world's greatest rainforest to a vast expanse of grass and bush. The latest science shows that the Amazon rainforest has recently changed from storing more carbon than it emits each year to emitting more carbon than it stores.

Much is still uncertain. 'The problem is we don't know how to reliably predict these abrupt changes,' says ocean scientist Andrew Shepherd. African rainforests continue to grow fast and accumulate carbon. But the new Amazon observations have forced climate modellers to increase their forecasts of future warming if we do not urgently rein in the accelerating trends. Richard Betts, at the UK Met Office Hadley Centre, says Gatti's findings suggested that rainforest decline could result in 40 per cent more carbon dioxide in the air by 2100 than has previously been supposed. Which means potentially five degrees of warming rather than three degrees. Some call it 'hothouse Earth'.

Such warnings should not make us fatalistic. They should gird us for action. To have the best chance of warding off tipping points, we have to keep long term global warming below 1.5 degrees. It is likely we will now pass 1.5 degrees in the short term but urgent cuts this decade give us a fighting chance of returning and stabilising below 1.5 degrees. What does that mean in practice?

Since the industrial revolution, we have dug and drilled from the ground and dumped into the atmosphere around 3 trillion tonnes of carbon dioxide. To limit long term warming to 1.5 degrees, we can only afford to emit around 10 per cent more, another 330 billion tonnes. That is about eight years' worth, at the current rate. Obviously, time is short. Realistically, say researchers, to have a chance of meeting that target we need to at least halve global emissions by 2030, halve them again by 2040, and reach net zero by 2050. That requires annual cuts in emissions of 7–10 per cent from now on. This is 'physically, technically and economically feasible', a group of leading climate scientists, headed by veteran Bill Hare, are now saying. 'But political action will determine whether it actually happens . . . The next ten years are crucial.'

This will require many actions. We will discuss some of them in later chapters. But top of the list is coal. Coal is our dirtiest fuel, and on its own produces a third of our emissions. Clearly we should not be building any more coal-fired power stations anywhere in the world. More than that, Fatih Birol, head of the International Energy Agency, says there must be an immediate global programme to shut all existing coal-burning plants, no matter how recently they were completed.

The bottom line is that the quicker we stop adding to carbon dioxide in the air, the greater the chance of getting a 'stable' climate back. Any delay – any failure to keep long term warming below 1.5 degrees – could see us passing tipping points in the climate system, beyond which there will be no way back.

Waste: Fouling our nest

Nature is an extraordinary recycling machine, in which life makes and sustains life. It achieves this because nature knows no waste. The product of one natural process is fuel or feedstuff for the next. Everything is recycled. Nature even erodes rocks to create soil, and recycles chemicals from the atmosphere to maintain the air that living things need to breathe. Once, when all our resources came directly from nature, we humans ran our lives like that too. Forest tribes reused everything. Farmers dug their waste back into the soil. The floors of many rainforests are covered in soils rich in kitchen waste, charcoal and human excreta left behind by past societies, for whom recycling was second nature. Even cities, right up until the late nineteenth century, collected their human excrement to put back on to 'sewage farms'.

But in our modern, throwaway, industrialised world, where we use manufactured materials that no longer degrade, we have forgotten about that. Most things are used once and then discarded. Often they never get used at all. A third of the food we produce is wasted. Most of our chemical fertiliser ends up flowing off fields, down rivers to the oceans, where rather than fertilising nature, it often kills it. Human-made industrial toxins and pesticides that nature has no use for accumulate in the world's oceans and even the Arctic. And solid waste – our billions of tonnes of discarded 'stuff' – ends up on street corners, in landfills, or joining the waste streams heading out to sea. What was once wind-blown litter has become a global crisis of microplastics, filling the oceans and invading the bodies of their inhabitants.

Our waste is just that – a crazy waste of resources that could often be put to productive use. And the deeper you delve into your waste bin, the crazier it looks. People worry whether the world can feed the extra 2 or 3 billion people we can expect by the end of this century. Well, the good news is we already produce enough to feed everyone; the bad news is that much of it goes in the bin. There is food enough for all, if we cut out the waste.

Fast fashion is another notorious culprit. Cotton is one of the world's thirstiest crops, and often requires huge amounts of pesticides. Yet we throw away 92 million tonnes of clothes each year, which works out at about 12 kilograms each. Rich countries are worst. The average American throws away 37 kilograms of clothes annually – and presumably buys another 37 kilograms to replace them. Globally, only around 12 per cent of the material in discarded clothes is recycled in any way – a much lower rate than for paper, glass or even many plastics. Most clothes sent for recycling end up not being worn, but as industrial cleaning rags or stuffed in mattresses and car seats.

• • •

Plastics are the latest waste plague. The volumes beggar belief. Humanity manages to flush away or bin 11 billion polyester and polypropylene wet wipes every year. That, by one estimate, makes a pile as tall as the Eiffel Tower every day, mostly smeared with removed make-up. They are major ingredients in the fatbergs heavier than trucks that occasionally block entire sewer systems.

We didn't have mass-produced plastics until 1950. But now the world produces some 400 million tonnes each year. In total, there are now around 5 billion tonnes of plastics littered across the landscape and floating in the ocean, exciting our anger with horrendous pictures of sea creatures snagged or choking on six-pack rings or fishing tackle. That is almost a tonne from each of us.

Plastics are the ultimate throwaway materials. And because most rot only very slowly, they stick around for centuries, slowly breaking down into ever smaller fragments – to microplastics and eventually nanoplastics. These tiny fragments get everywhere, including into you. Every year you will probably eat enough minute pieces of nanoplastic to make a credit card. You breathe them in too. The tiniest bits can cross into your bloodstream, cells and brain. Does that make them dangerous to our health? Maybe, or maybe not. We simply don't know.

A surprising amount of microplastics is floating through the air. The small particles can float on the winds for several days. But whether floating down rivers or through the air, much of it ends up, like so much of our long-lived detritus, in the oceans. We can see this from the scummy rings of plastics on remote beaches. Ascension Island in the South Atlantic is a thousand kilometres from the nearest land, but that doesn't stop its shoreline being covered in plastic trash, most of it the packaging of products not sold on the island. Scientists have mapped ocean currents using the plastic released by containers that fall off ships, such as a consignment of rubber ducks lost in the North Pacific in 1992 that eventually beached everywhere from Alaska to Australia and Scotland, and plastic casings of ink-jet cartridges lost near the Azores in the North Atlantic that were turning up on the Arctic shores of north Norway four years later.

The most dramatic accumulations are in the five great ocean garbage patches, where circling ocean currents known as gyres collect and trap floating waste. The biggest and most famous is the Great Pacific Garbage Patch, between California and Hawaii. It holds 80,000 tonnes of plastic, much of it more than two decades old. Others are in the North and South Atlantic, the Indian Ocean and the South Pacific. It is a crazy ending for crazy products. But all too predictable in a world where, as Tamara Galloway at the University of Exeter told *Nature* magazine, 'we produce things that last for 500 years, and then use them for 20 minutes'. For the remaining 499 years, 364 days, 23 hours and 40 minutes, they are waste.

• • •

Metals get recycled at rates that generally reflect economics rather than ecology. Most aluminium is recycled, because it takes a lot of energy to smelt more from the bauxite ore. The electricity needed to make aluminium for one drinks can could run a television for three hours. As a result, three-quarters of the billion tonnes of aluminium produced since manufacture began in the 1880s remains in use somewhere. Your drinks can – or the aircraft you fly in – may be made from metal that made an early bicycle, or a canteen that held food for soldiers in the trenches in the First World War. Similarly, around 80 per cent of the copper ever mined is still in use.

The fastest growing category of our garbage is electronic waste: kettles and toasters, freezers and washing machines, mobile phones and TVs, thermostats and microwaves. Worldwide, we throw away 53 million tonnes of this stuff a year, or over 7 kilograms per head. It is not as voluminous as plastics, but it is often much more toxic.

Less than a fifth of our e-waste gets recycled, much of it in dangerous circumstances. Millions of poor people in poor countries, with few alternative means of scratching a living, dismantle container-loads of abandoned equipment to extract small amounts of valuable metals, from gold and silver to copper and aluminium. Whole city neighbourhoods specialise. In the Indian capital New Delhi, Mandoli suburb is devoted to dismantling computers – many of them illegally exported from Western countries. Migrant child workers with persistent coughs labour over giant fuming vats that boil circuit boards in acid to liberate copper to supply nearby wire factories.

Even more dangerous is the global backstreet business of extracting lead from used car batteries to make new batteries. It may be recycling, but it is also often lethal.

Kenyan mother Phyllis Omido thought her baby had malaria until doctors discovered the dangerous levels of lead in her breast milk. She worked as a business manager at a battery recycling plant in a densely populated suburb in Mombasa. It was poisoning an entire community. When her bosses refused to clean up, Phyllis quit her job to campaign to get the recycling plant closed. It was a tough task. The government's pollution police, who had originally licensed the plant, said she was lying about the problems. Many of her former colleagues feared losing their jobs. She was attacked at night by armed men, and went into hiding.

On the other side of Africa, in Dakar, the capital of Senegal, eighteen children died in just three months from a brain disease called encephalopathy. Doctors found it was caused by lead pollution from another local battery recycling plant. Fumes from the plant's smelter had left so much lead in the ground around the plant that locals scraped up the soil to sieve out the metal for sale. They stored the soil in their homes. Their children played in it. Hundreds were poisoned.

Lead is well known as a neurotoxin. There are no known safe levels. Once it was widely used as an additive in petrol, until the hazards were exposed and it was banned. Today, around 85 per cent of the metal goes into the batteries in almost all of the world's 1.4 billion motor vehicles. When the batteries run down, about 6 million tonnes of lead is collected for reuse every year. One study estimates there could be 90,000 lead recycling plants worldwide. 'But we don't know. The more you look the more you find,' said

Richard Fuller, head of Pure Earth, an NGO that has researched the issue.

The industry boasts that lead batteries are 'the world's most recycled consumer product'. But at least half of this business begins with backyard battery breakers and slum smelting plants. From Vietnamese villages to Chinese megacities, from Roma camps in Eastern Europe to forest clearings in Bangladesh, it is one of the most serious – and least known – environmental hazards for millions of children.

Fuller doesn't want to stop recycling. What is needed, he says, is to make manufacturers set up a proper organised recycling industry that collects all used batteries, sidelines the backstreet operators, and cleans up the contaminated sites. Organised collection is routine in Europe and North America – though a disturbing number of US batteries are trucked south to Mexico for cheap recycling. Countries such as China and Brazil are now acting to control the industry. But elsewhere the lethal backstreet industry continues. Still, in Kenya, Phyllis eventually won her campaign in Mombasa. Courts shut down the plant that poisoned her baby, ordered $12 million in compensation for the families of children that died – and initiated a clean-up. It is a start.

• • •

Increasingly, our recklessly discarded waste is disrupting basic planetary processes. Our CFCs almost burned up the ozone layer. Our carbon dioxide is messing with the atmosphere's carbon cycle, our planet's thermostat. And now we are trashing the nitrogen cycle, one of the planet's most important ways of recycling the materials essential to life on Earth.

Nitrogen is a vital component in soil nutrients, essential for all plants. Living things use it as they grow and leave it behind for the next generation, in rotting vegetation, corpses and animal dung. But humans have disrupted this constant recycling. We leave little to rot in our fields because we harvest and eat the crops. We don't return our faeces to the land any more, instead pouring it down drains and into rivers and the ocean.

To make up the difference – and because of our soaring demand for food – we add increasing amounts of nitrogen fertiliser to our fields. Once we obtained nitrogen by extracting bird droppings on remote islands round the world. Then a century ago, as demand exceeded supply, German inventors came up with the Haber-Bosch process, which captures nitrogen from the air and converts it into chemicals that we can pour on our fields.

By boosting crop yields, the Haber-Bosch process has doubled the world's farmlands' capacity to feed us. It sustained the population explosion through the twentieth century. We grow ever more reliant on it. Most of the human-made nitrogen fertiliser ever produced has been applied to fields in the past thirty years. But we also waste it recklessly. Of the 120 million tonnes we spread each year, only 40 per cent fertilises crops. In China, the figure is just 25 per cent.

The rest washes off the fields into drainage ditches and

rivers, and ultimately the oceans. There it joins the 70 million tonnes of nitrogen from our sewage. The result is an ecological disaster. Because while nitrogen is essential for living things to grow, too much of it in water creates plagues of algae that consume all the oxygen in the water. Everything else chokes to death. In many parts of the world, ecosystems are under siege from excess nitrogen. In China, it kills fish in huge numbers from the Yellow River in the north to the Pearl River in the south. Blooms of toxic algae form in a third of the country's lakes. Massive red tides spread from its river estuaries across the East China Sea. Worldwide, there are more than 400 ocean 'dead zones' produced by nitrogen pollution that cover an area the size of the United Kingdom.

While the world is gradually getting more efficient in the way it uses most resources, we are still becoming ever less efficient in how we use nitrogen. It makes no sense to pour our nitrogen-rich sewage into the oceans, when applying it to soils could replace up to half of the world's chemical fertiliser. It makes even less sense to waste most of that chemical fertiliser by allowing it to run off the land and choke natural ecosystems. But we have yet to figure out a way of doing things better.

When Swedish environmental scientist Johan Rockström convened a group of his fellows to draw up a list of 'planet-ary boundaries' – his term for the edge of the safe space before we risk hitting tipping points in natural systems – they listed three where we are right on the edge. One was carbon dioxide emissions causing climate change; a second was mass extinctions and the loss of nature's biodiversity;

the third was saturating the planet's ecosystems with nitrogen.

The tragedy of our mishandling of nitrogen shows vividly the sheer wastefulness with which we use the world's resources and pollute the environment. But it is also our potential salvation. For, if we waste more than half the nitrogen we pour on our fields, then we can stop wasting it. Similarly, we could start seeing our sewage as a resource, and make good use of it. We cannot 'flush and forget' our nitrogen any more.

And that applies to all our wastes. As we said at the start, nature knows no waste. For nature, everything is a resource. We must take the same approach. In Chinese philosophy, yin and yang are the opposite and seemingly contradictory sides of the same thing. Our garbage is the yin and yang of environmentalism. Looked at one way it is waste; looked at another way it is a valuable resource waiting to be recognised – and used.

3

Reasons for Hope

Any planetary doctor would say the Earth is in a critical condition. One recent scientific review of the state of nature and its impact on our lives said we face 'a ghastly future of mass extinction, declining health and climate disruption'. The planet's life-support systems are on the cusp. The five crises – climate, waste, nature, air and ocean – are now exacerbating each other.

Forest loss and oceans emptying of everything from phytoplankton to shellfish both make the climate crisis worse. Meanwhile, the climate crisis triggers wildfires that rip through forests, bleaches coral reefs that nurture fisheries, floods coastal wetlands, sends species to extinction, worsens city smogs, and releases methane from permafrost, risking runaway warming. Air pollution kills forests, eats at the ozone layer that protects all nature, leaves tundra radioactive, and kills all too many humans as well. Our waste poisons rivers and worsens the ocean crisis, while oceans without fish can push us to clear more forests to grow crops.

Everywhere we look, the ratchet of ecological destruction is turning, tipping points are approached and runaway effects loom. Few on the front line of research into any of these issues doubt that the coming decade is a moment of choice for humanity. Planet Earth needs intensive care now.

Yet now is also the moment of hope. For at last we recognise where we stand, and the crises we have created. We know natural forests have to be protected and restored, and that wetlands, grasslands and marine ecosystems need the same care. We know that we have to end our addiction to fossil fuels now, and instead keep the carbon in the ground, in ocean sediments and within ecosystems. We know the health of our children and much of our wildlife depends on cleaner air. We know that our consumer culture and scorched earth approach to ecosystems have both run out of room. We now know without any doubt what we must do, and with that knowledge comes the freedom and impetus to act.

Humanity has never before had such shared knowledge, never before had the ability to collaborate globally at such scale, never before had the extraordinary opportunity to create our own future.

Ecologists have been saying some of these things for a long time. It is half a century since the Club of Rome compiled *The Limits to Growth*; since Paul Ehrlich wrote *The Population Bomb*; since Rachel Carson told us about the *Silent Spring*; since E. F. Schumacher wrote *Small is Beautiful*; since the *Ecologist* magazine drafted *A Blueprint for Survival*; since James Lovelock came up with his theory of Gaia; and since Barbara Ward and René Dubos co-wrote

Only One Earth in the run-up to the first Earth Summit in 1972.

But now in 2021 even bankers and insurance bosses, CEOs of oil companies and US presidents say the same. 'The build-up of greenhouse gas led to changes in our climate that, if left unaddressed, present existential consequences for the whole planet,' said Mark Carney, governor of the Bank of England until 2020. US president Joe Biden has called the 2020s the 'decisive decade' for fighting climate change, and announced a $2 trillion plan to do just that. The chief of the world's largest financial assets manager, Larry Fink at BlackRock, warned of 'compelling' risks posed by climate change that were forcing 'a reassessment of core assumptions of modern finance'. Coal companies were going bankrupt, and oil companies such as BP claim they are getting out of oil.

And it is not just words. Or not always. In places, we are turning the tide. Interestingly, the story starts in Britain. The country began the industrial revolution more than two centuries ago and now could just be in the vanguard of a new revolution, to make industrialisation safe for the planet and its near 8 billion inhabitants.

In the early nineteenth century, at the height of its industrial dominance, the UK topped the world in manufacturing; it mined 80 per cent of the world's coal and most of its iron. It no longer has that place in the world, but, like most countries, its wealth continues to grow. Yet the country's consumption of materials such as food, textiles, construction materials, metals and fossil fuels peaked in 2001, according to government statistics: today this figure is less than a third of what it was twenty years ago – down from 15 tonnes of stuff per head per year to 8.3 tonnes by 2019. The UK now pollutes less, too. Its greenhouse gas emissions are down 38 per cent since 1990 – though this does not count emissions created in producing products imported from overseas.

The UK is not alone. Other rich developed countries have also apparently arrived at what environmental analyst Chris Goodall calls 'peak stuff'. Europeans consume 18 per cent less than they did a decade ago. Most rich countries, including the United States, are using less energy, both in industry and within households, as appliances become more efficient. We drive less too. A generation ago, getting a car was a passport to adulthood. Today, owning your own wheels has lost its social cachet. Many young people don't have a driver's licence, even in the United States, once the home of car culture. Goodall sums up: 'Water use is down, travel and car ownership are down, metals and paper use down, cement use down, and meat eating is falling.'

There are plenty of potential reasons for this. To some extent it is because the West imports what it once manufactured or grew – often from China. But equally, there is often reduced national spending on infrastructure, so less concrete is poured and asphalt laid. That may be temporary, but there is also greater manufacturing efficiency, with a jump of 60 per cent in the economic productivity of each tonne of stuff produced.

It is also about technology. One optical fibre can do the work of a thousand copper phone wires. Digital computing, modern telecommunications and automated manufacturing, robot farming, 3D printing and many other technologies offer the potential to do things many times more efficiently, less wastefully, and less destructively for the planet.

But it is also true that many citizens have personally stopped consuming more stuff. Some of this is again better technology. A smartphone does all the things that used to require a radio, a camera, a newspaper, an alarm clock, a CD player, a torch and more. Some is about lifestyle. We drive less and bike more. We are starting to eat less meat and more vegetables; have more gym time and less tobacco-smoking time. We fill our homes with less stuff and instead spend our money on experiences. Some experiences, like foreign travel, use a lot of energy. But others, like yoga classes, fitness trainers and nice restaurants, do not. And if we spend all day communicating online, rather than in our cars travelling to see people, that is another gain for the planet.

Some economists believe that something big is happening here. We have, they say, begun to 'dematerialise' our world; this is perhaps our tipping point towards a good Anthropocene. Jesse Ausubel of the Rockefeller University in New York City says that 'our economy no longer advances in tandem with exploitation of land, forests, water and minerals'. We can decouple growth from resource exploitation and pollution. We can live better lives with less.

Others point out that the world as a whole is still going in the wrong direction. A few countries may have reached peak stuff, but global steel production has almost doubled since 2000, and cement use has soared 160 per cent. Julian Allwood of the University of Cambridge says even if the rich world is sated with stuff, 'hundreds of millions of people are still at the very beginning of that global consumption rise'. Even green energy is double-edged, they say. The planned growth of electric cars, solar panels and wind turbines could trigger a fourfold increase in demand for nickel and lithium for batteries, according to the International Energy Agency.

The question for the twenty-first century, and perhaps for the future of humanity, is this: Which version of modernity is set to triumph on our increasingly crowded planet? Is it the orgy of construction, manufacturing and consumption – trashing the water and air, destabilising the climate, and degrading the ecosystems on which our global civilisation ultimately relies – instigated by the West and now ripping up much of the rest of the world too? Or is it the dematerialising society now glimpsed in some parts of the industrialised world, where innovations in technologies allow standards of living to remain high while consumption of material goods falls to a level our planet could sustain? Could technology, which in the past has caused so many of the planet's problems, now provide some of the solutions? In particular, could it give those billions of people in developing nations who still crave what the rich world has a shortcut that will bypass the environmentally ruinous road to riches taken by their predecessors?

• • •

We know that when we put our minds to undoing the damage, we can achieve a lot. Even the natural world shows this. Thanks to our conservation efforts, some of our favourite endangered species are coming back from the brink. Take tigers. They have lost 95 per cent of their historical range. People kill them for traditional Asian medicines, for their fur and to stop them hunting livestock or attacking children. But their numbers are bouncing back, nonetheless. In 2010 there were as few as 3,200 tigers in the wild, a collapse from the 100,000 a century ago. But the corner may have been turned. We could be back above 4,000, with more than 3,000 in India alone. Numbers in Nepal have doubled. In Russia, Bhutan and China, increased sightings suggest conservation efforts are also working.

Likewise, the giant panda. From a low point of about 1,100 in the 1980s, they may be back to 2,000 now. Recently, Chinese conservation officials have announced that they no longer consider giant pandas in China an endangered species. American bison on the Great Plains were hunted from 60 million to fewer than 600 a century ago. Today they are back to 60,000, roaming national parks and other unfenced areas set aside for them. European bison too are creeping back – to 7,500 at last count. In East Africa, constant protection has allowed mountain gorillas to double their numbers to more than a thousand today. Loggerhead and green turtles are recovering in the Mediterranean since hunting was banned and nesting beaches were protected. Since China banned the ivory trade in 2017, elephant poaching in Africa has dropped sharply.

It is also encouraging to know that, if left alone, nature

often thrives almost unobserved. While some great animal migrations have disappeared, others still thrive. Millions of white-eared kob, a type of antelope, were discovered a few years back still racing out of and back into war-torn South Sudan across fenceless grasslands into Ethiopia, a journey of more than 800 kilometres. And the longest migration by any land animal on Earth is in rude health.

The great Porcupine caribou (American reindeer) herd moves annually some 2,400 kilometres from its winter home in northern Canada to summer calving grounds in North Slope, Alaska, on the shores of the Arctic Ocean. Their migration coincides with the emergence of the lush grasses in North Slope, providing ample food for the new young. Fears that climate change could mess up the timing and leave them short of food for their young have so far proved false. Their resilience may have limits, and some other herds are in decline. But the size of the Porcupine herd reached a new peak in 2017, when an estimated 230,000 animals made the journey, double the number two decades before.

None of this says nature is doing fine. What it does say is that nature has resilience and adaptability. We need to enable and enhance those traits, and above all to give nature room.

Ecosystems can bounce back too. Costa Rica has almost doubled its forest cover in the past three decades, largely through natural regeneration. Once, the Central American country was a byword for deforestation. It cleared its trees to supply American burger chains. At one point, the country reportedly sold 60 per cent of its beef to Burger

King alone. Tree cover fell from 75 per cent to less than 30 per cent through the twentieth century, until its government got tired of being an environmental pariah, and started paying landowners to protect and restore forests. Now forest cover is back above 50 per cent, most of it lush rainforest supporting a booming eco-tourist business.

Other developing countries are following in Costa Rica's footsteps, often as farmers give up their land to move to cities. South Korea has doubled its forest cover. Nepal is 20 per cent up. Chile, El Salvador, India, Thailand, Panama and Thailand all have more trees. Sometimes these are plantations rather than naturally regenerated forests. Still, there are more trees in the world today than a generation ago, and more than 90 per cent are, if not pristine, then substantially natural forests.

Without going overboard, there are other good signs that we can and sometimes do make good the damage we have done. Some of the nastiest industrial poisons and pesticides, such as PCBs and DDT (dichloro-diphenyl-trichloroethane), may still be circulating in the environment, but at least production and use is banned in most countries. Thirty years ago, two of our biggest global environmental concerns were the poisoning of our children from lead in petrol fumes and the damage CFCs were doing to the ozone layer. Both were the result of inventions by an American chemist called Thomas Midgley, who was feted in his day as a great industrial innovator. Environmental historian John McNeill wrote that Midgley had 'more impact on the atmosphere than any other single organism in Earth's history'.

The good news is that we fixed his mistakes. The hope

is that in the twenty-first century our great inventors can have similar great impacts – but of the right kind. So let's step back to make two statements that give grounds for optimism that we can in the next decade stop doing harm and instead prepare the ground for a good Anthropocene.

The first ground for optimism is that, especially if we give it a helping hand, nature has great powers of recovery. Not always, but often. We can clean up our mess, reintroduce displaced species, plant trees and so on. But our helping hand may often involve no more than ceasing to 'do bad'. If we put away the chainsaw and the plough, then forests and grasslands will usually regrow. If we tear down dams, then many rivers and wetlands will recover. If we stop wrecking reefs with fishing nets and dynamite, ships' anchors and cyanide, then coral will often regrow. All these repair mechanisms can be cut short by climate change. So we have no alternative but to fix that as fast as possible, or all bets are off in 'hothouse Earth'. But nature will do its damnedest.

In Japan, they still marvel at what happened in the aftermath of humanity's single worst moment of mass destruction. The atomic bomb, dropped on Hiroshima in 1945 in the closing days of the Second World War, left 140,000 dead. Today, the city has a museum to describe what happened. As visitors leave, they are approached by a museum guide who bows and says: 'Please remember that within a month of the bomb, the grass across Hiroshima started to grow again.' Behind her, archive photographs show how, even as irradiated victims were buried in mass graves, plants were proliferating along riverbanks and bursting through fissures in the roads made by the blast. It was, the guide says, 'a message of hope'. Nature was renewing itself.

There are limits, of course. We cannot assume that nature will always make things right. Nor that any new world that nature creates will be to our liking. Nature won't do our bidding. It has its own agenda. But that is its strength, and we should applaud and nurture its capacity for adaptation and reinvention.

The second ground for optimism is that humans can change their ways. We do it constantly, and often for the better. Nobody a generation ago would have imagined that many of us would now live in a world without a constant fog of tobacco smoke, where drink-driving is banned and where same-sex relationships are celebrated in many places. But we do, and this is because we have chosen to do so. And sometimes we change quite suddenly for, like nature, we too have 'tipping points'.

Through technical innovation and societal change, we can heal our planet. In some limited ways, we already are. Recycling some of our household waste and buying products with eco-labels is now normal behaviour for many of us. Protecting rather than hunting wildlife is a new social norm for many. Developing renewable energy has transformed our energy systems. Solar energy was once so expensive it was only used in spacecraft. Today, it is often as cheap as coal.

Now the need is to do much more, and go much faster. We need to celebrate the innovation, embrace nature's powers of recovery, and find the personal and social tipping points that will allow us to be good citizens in a good Anthropocene.

Listen to the words of Johan Rockström. He has been our 'conscience in chief', as the formulator a decade ago of the nine planetary boundaries. 'We are running out of time,' he says, in an interview with one of the authors for *New Scientist*. 'For climate we have to cut emissions by half over the next ten years, or we will be well into a high-risk zone. We are beyond our safe operating space on planet Earth.' Beyond the boundaries, he says, 'the changes could be abrupt and irreversible. We don't know where things may end up.'

But even he, with his detailed knowledge of how close to disaster we are, is ultimately optimistic. 'Over the past ten years, we have seen an exponential rise in sustainable energy solutions, with solar photovoltaics, wind power and electric mobility all much more economically competitive.' A motorbike fan, he adds that 'even Harley-Davidson is going electric'. We may, he says, 'be at a positive tipping point in humanity's response to the threats, entering a new era, a renaissance in which sustainability is essential to the success of businesses. The question now is whether the change is happening fast enough, and we don't know the answer to that yet.'

Achieving those positive tipping points and supercharging that change is the purpose of The Earthshot Prize. It is founded in a belief that we the people – as investors and inventors, producers and consumers, governments and citizens – need not be planetary pariahs. That we can be planetary saviours.

PART 2

FIXING OUR PLANET

In Part 1 we looked at the scale of the problems our species has created – for planet Earth, for nature and for ourselves. We suggested some guarded grounds for optimism, seeking out signs that we are not on a road destined to end in disaster. That may still happen. But it need not. Even as our numbers rise towards 8 billion, we have choices to make. The future remains in our hands. We have the chance, perhaps a last chance, to fix things. As a former British government chief scientist, David King, puts it in an interview with one of the authors: 'Time is no longer on our side. What we do over the next ten years will determine the future of humanity for the next 10,000 years.' We can, if we choose, have a good Anthropocene. But it requires a change of perspective. One where we learn to respect our place as part of nature, rather than removed from it, and where we take responsible charge of the great forces that govern the habitability of our world.

Now we come to the crux of the matter: how to do it. We know what to do in outline, but we don't yet know the details. Unleashing the innovations that will enable us to discover those details is the ambition of The Earthshot Prize, and what the following chapters are about. They don't offer a road map to our ecological redemption, but they do shine a guiding light on the way ahead. They emphasise human endeavour: how individuals can make a difference through making technical innovations that allow us to do more with less disruption, through leading social transformations to reduce our demands on the planet, and by harnessing nature's powers of recovery. They stress the urgent need to harness these ideas and scale up the innovations and inventions to make real and big differences in the coming decade. Nothing else will do.

We look at the five Earthshots in turn. Each chapter is introduced by a member of the Prize Council with a personal connection to that particular Earthshot.

Protect and Restore Nature

T hroughout my lifetime I've been lucky to witness more wonders of the natural world than I could have ever dreamed of. I've seen the greatest migrations, travelled alongside the largest animals that ever lived, encountered our closest relatives, and been dazzled by the beauty of nature. I've felt the cold of the poles and the heat of the great deserts. And I've been fortunate enough to share many of these experiences with audiences around the world.

In recent years I've also felt a responsibility to consider with audiences how much of our natural world we have lost and the consequences that will follow if we don't reverse this trend. But there are also stories of hope and these are just as important to tell. For nature can recover – sometimes with our help and sometimes when simply left alone.

There are many such examples around the world but one I've visited recently is much closer to home: the Knepp Estate, in the South of England. Fifteen *years ago their land was a typical modern arable and dairy farm, with field after field of crops fed with fertilisers and sprayed with pesticides. Much of the wild had gone.*

But then the owners made the decision to change their approach to managing the land. They decided to work with nature, rather than against it. They stopped the spraying and fertilising, they took down the fences between the fields, they stocked the land with a mix of livestock that resembled the animals that once roamed wild in this part of the world. In short, they gave the natural world the chance to rebuild.

And rebuild it did – from the ground up. With the return of the insects, the wildflowers and weeds and the more natural ruminant animals, the soil itself changed for the better. And the soil fed those plants and crops better, and the insects that ate them fed the larger animals that came back to the area. The foundation of all life starts with the little things, things so small we need a microscope to see them. It is now one of the most diverse patches of land in the South of England. Given half a chance, nature found its way back, and it did so quickly.

There is no single solution to restoring biodiversity – Knepp is simply one example. Every patch of land will require its own unique approach and each will come with trade-offs. But there does seem to be an attitude common to most successful restoration projects. Those that embark on them take the time to understand nature and find the ways to work with natural processes. In doing so they discover the ways they can still use the ecosystem for food, timber, recreation or simply to live in, while allowing nature to recover.

When we talk about re-wilding, or of restoring biodiversity to an area, it doesn't necessarily mean that we can turn things back to the way they used to be – nor that we have to. In England, for example, in lieu of conjuring up any extinct wild Aurochs to graze the land, the right sort of domestic cattle can serve as a good replacement. These cattle, along with ponies and deer, crop the grasses and browse bushes and trees, encouraging diverse meadows and natural thickets. By doing this in a managed way, they can allow the free-range farm animals to play the part of their wild cousins.

The planet as a whole is now a lot less wild than it was when I began my travels seventy years ago. By reducing the variety of nature, we have weakened the living world, leaving it less stable and more fragile. But as many successful restoration projects show, we can *restore the natural world; we just need to choose to do so. It's not too late. We can start to make more space for nature.*

We can learn from each other, and from the many people trying to restore nature to their own patches across the world. If we choose to spend the next ten years perfecting all the new ideas we have seen, and searching for many others, just imagine what might be possible. A world in which landowners gain more from building diverse, wild habitats than tearing them down. A world in which we can work with, rather than against, natural cycles. And a world in which we become so efficient at providing for ourselves that we can spare enough space for the rest of life on Earth.

Tomorrow's world could be more diverse, more stable, more wild, if we start making the right choices today.

Sir David Attenborough, 2021

For many centuries, the call of the wild for many people has meant the howl of the wolf. The animal and its cry combined our sense of both the wonder and terror of nature. Now in Yellowstone, America's first national park, scientists are learning afresh how these great hunters are not just symbols that resonate in our imaginations, but how they are central to the functioning of their ecosystems.

Yellowstone, a volcanic plateau in the Rocky Mountains of Wyoming, was declared a park in 1872. Back then, the idea was not to preserve a pristine ecosystem but to create a wild playground for colonising Europeans. So Native Americans were expelled. The lakes were stocked with non-native fish. To increase numbers of elk and other game for hunters, their main natural predators, the wolves, were exterminated. It took time, but the last grey wolf was shot in Yellowstone in 1926.

For decades afterwards, Yellowstone had no top predator. That messed badly with the rest of the wilderness, says wolf specialist Kira Cassidy of the Yellowstone Wolf Project. 'Predators are extremely important to ecosystems. Without them, ecosystems go into hibernation, almost suspended in time.' But nobody knew what they were missing until the decision was taken in 1995 to launch a wolf reintroduction programme. Thirty-one wolves were released.

It had initially been an experiment aimed at reducing the number of elk, which were running out of control. But the effect was much more profound. 'When the wolves were returned, it was like someone just pushed a play button, and everything started to work properly again,' says Cassidy, who has been monitoring the transformation ever since, following the wolves night and day, in person and with radio tags.

In the decades without wolves, elk numbers had soared. They had grazed vegetation so low that trees such as willow and aspen rarely grew more than knee-high. But with the wolves hunting them, elk numbers fell from 15,000 to around 4,000, and the vegetation swiftly began to regenerate. That much had been anticipated. But there was much more. Other animals, such as the much bigger moose, liked the resurgent vegetation and were rarely hunted by the wolves. So their numbers grew. Songbirds returned to nest in willows now tall enough to be out of reach of the elk. Meanwhile, as the resurgent vegetation stabilised riverbanks and provided wood for their dam building, beavers were back in business. They increased from one colony to more than thirty. The beaver-built dams raised water tables, created pools that encouraged the return of native fish, and formed boggy ground where yet more aspen and willow could grow.

Meanwhile, the hunting wolves left behind animal carcasses that attracted scavengers for the skin, bone and anything else the wolves did not eat. 'We were seeing coyotes and foxes, bald eagles, ravens, even grizzlies and black bears coming in and getting food from the wolf kill,' says Cassidy. Almost every species in the park benefited from the return of the apex predator. 'It has been a domino effect. In only a decade we were seeing massive changes.'

Cassidy says the story of Yellowstone's wolves shows both what is often lost when top predators are removed from an ecosystem, and the dramatic upsurge of nature that can follow their reintroduction. 'Yellowstone has been called the most successful conservation story of the twentieth century. It shows how biodiversity can come back,' she says. The lesson needs to be learned elsewhere. 'Many different ecosystems across the world are not operating the way they should be, just because their predators have been killed off.' Crocodiles and lions, cougars and leopards have all been persecuted till they are close to extinction in many of their natural habitats. We need to recognise the extent of the loss. But we also need to hear the call of the wild and bring them back. For once restored, says Cassidy, the return of a whole variety of nature can be 'shockingly quick'.

· · ·

Predators are key to many ecosystems. They are like the lead instrument in an orchestra. But other species matter too for the music to play. It is nature's full repertoire – from the carnivores, through herbivores and plants, right down to the smallest bugs, flies and fungi – that ultimately matters. Ecosystems, and their resilience in the face of change, are founded on what we now call 'biodiversity'.

The word 'biodiversity' is a relatively new one. It was invented by conservation scientists in the 1980s. But it describes well what is at stake: the diversity of nature. Nature's orchestra comprises trillions of individuals of millions of species, all evolved from each other and working together in a complex web of life. Plants need insect pollinators and herbivorous animal seed carriers if they are to reproduce. Giant whales need plankton to eat, but also stir up the oceans, providing nutrition for the plankton. Carnivorous predators need herbivores, and by eating them drive entire ecosystems.

Many species evolved in tandem with other species that live with them, something biologists call 'co-evolution'. Even so, the web is constantly changing, with species coming and going. Though the key links in the chain don't change much, ecosystems are constantly evolving. Ecologists say it is the complex interplay of stability and dynamism, cooperation and competition, fragility and resilience that allows nature, and in turn humanity, to thrive. They say we have done much harm to this complex web, but that the inherent dynamism of nature gives many ecosystems an excellent chance of recovering – given the chance.

The return of the wolves is driving regeneration in Yellowstone. Something similar is being attempted on an even grander scale not far from there, on the Great Plains in Montana. Here, the star attraction is the return of the buffalo. Teeming herds of these American bison once numbered in their tens of millions. They ruled the grassy plains. Native Americans lived on their bounty, until the arrival of European hunters, who reduced the great beasts to within a few hundred of total extinction by the mid nineteenth century. But a few survived, and today they are back, mostly bred from a surviving herd in Yellowstone. Thousands once more roam national parks, and thousands more live on private bison ranches, such as one owned by media mogul Ted Turner that produces bison burgers for his chain of restaurants.

For ecologists, the real prize is not more bison on ranches, but the return of the ecosystems they once grazed. In northern Montana, a land trust set up by WWF is piecing together an American Prairie Reserve, buying up cattle ranches south of the railroad city of Malta. Wealthy philanthropists, including the Mars family of chocolate fame, are bankrolling the enterprise. As of 2021 it had 170,000 hectares of former ranch land – a seventh of the way to its eventual target.

Fences are being torn down, the old ranches are linking up with unfenced public areas, and the grasslands are being restored to recreate what some like to call an 'American Serengeti', comprising wolves and cougars, grizzly bears and bighorn sheep, elk and prairie dogs, birds and snakes, amphibians and above all the bison. In this part of the world, at least, a corner has been turned. With human assistance, nature is being restored. But there is a huge distance to go.

• • •

The scale of our impact on nature is breathtaking. As we saw in Part 1, humans have turned almost half our grasslands into farms, and cleared more than 3 trillion trees. We have remade and plundered landscapes so much that the weight of all our inanimate stuff is now greater than the weight of living things, and 96 per cent of all mammals are either us or our livestock.

This ecological holocaust has in many places destroyed nature's ability, developed through evolution over billions of years, to control the planet's life-support systems – from the chemistry of the oceans to the biology of the soil and the physics of the climate. In the past century the carbon and nitrogen cycles central to these systems have been fundamentally altered. And as species disappear and biodiversity is diminished, so nature's ability to evolve and adapt especially to the speed of changes we are making is crippled.

As these systems break down, the planet faces a fundamental crisis that scientists are only now beginning to understand. They struggled to persuade policymakers that we all have an urgent duty to turn things round. Not just to halt the destruction, but to begin the repairs.

Still, it is starting to happen. The United Nations has set the 2020s as the Decade of Ecosystem Restoration, when the world must 'prevent, halt and reverse the degradation of ecosystems' on every continent and in every ocean. It is calling on governments and others to restore to nature a billion hectares of land – an area about the size of China. The restoration is not just about forests, though they get the most attention, but also about returning to nature 'farmlands, grasslands, mountains, peatlands, urban areas, freshwaters and oceans'. It requires ending human-caused extinctions and habitat loss, protecting biodiversity hot spots, and nurturing nature's resilience and dynamism.

To achieve this will require going far beyond conventional ideas about conservation as something carried out by governments, volunteers, non-governmental organisations and rich philanthropists on land set aside from human activity. All have a role, of course. But it also means finding ways to make banks, investors and business people value nature in their day-to-day decisions, by rewarding nature's protection and restoration, and penalising its destruction. Veteran environmentalist Amory Lovins says we need to rewrite the rule book of economics, so that we do 'capitalism as if nature mattered'. To find a new way of doing business as usual that does not continue the destruction.

Mark Carney, when governor of the Bank of England, said: 'The current system places great value on Amazon the company but no value on the Amazon rainforest – until it's cut down and turned into timber and so-called "productive land". It is that system we need to change.' There is growing recognition too that this will – and should – often involve returning control of natural lands to the indigenous peoples and local communities who know and nurture them best.

In the Serengeti plains of East Africa, the local Maasai cattle herders have long shared the wide open grasslands with elephants, giraffes, zebras, leopards, lions, buffalo, wildebeest, rhinos and more. Today, as they are hemmed in by farms and other economic developments, they are finding new ways to share their land with its iconic wildlife – by reducing their cattle-raising and inviting in tourists. Their form of 'capitalism as if nature mattered' gives cash value to the nature in their midst, and allows them to benefit from it. It is capitalism as if the Maasai mattered, too.

Perched on a clifftop above a waterhole frequented by elephants, leopards and giraffe, north-west of Mount Kenya, is Il Ngwesi, Kenya's only Maasai-owned and -run tourist lodge. It was built with local materials in the 1990s by young Maasai leader Kip Ole Polos and a team of warriors. It has a pool, terrace bar and spa, and is a marvellous base for exploring the bush, encountering its teeming wildlife, and visiting local villages. But it is also a blueprint for community conservation. The 5,000-strong Il Ngwesi clan run the place, own the land and take the profits. With nightly tourist rates of around £130 per head, the income is enough to send their children to school, their sick to hospital and their youth to university, as well as to pay for community guards to watch over their 6,600-hectare territory – and keep the tourists safe.

The Il Ngwesi people still have their cattle. 'Cattle are very important to the Maasai,' says Kip. 'To be a real man you have to have cattle. They are a symbol of wealth. We believe we came from heaven on a hide.' But now there is wealth in the wildlife too, he says. The Il Ngwesi logo is a warrior walking with a lion. 'Maasai warriors are known for killing lions, but we're not killing them any more. We are friends,' he says.

The lodge and the surrounding Maasai land show how community conservation can work. When Kip was a young man, the tribal land was running out of control. It was bad for people as well as wildlife. 'There was a lot of overgrazing, and poaching,' he remembers. Gangs of poachers, known locally as 'shifters', went around killing people as well as the wildlife. Migrating elephants rushed through the territory, knowing they were at risk. And his people stayed home.

'But that has really, really gone down now,' Kip says. The area is now formally designated as a conservancy, one of many being created across Kenya. A grazing plan agreed by the community with government protects wildlife and the bush. The Il Ngwesi now feel they are in control of their lands again, and the benefits are tangible. 'The trees have grown back, grass has come back, the wildlife have come back, birds have come back, endangered species like Grevy's zebras have come back,' he says.

Wild animals, cattle and the Maasai mingle happily together in a way not seen for decades. The elephants no longer rush through, Kip says. 'They know the exact areas where conservation is being practised. They have got to understand that it's safe, and people who are herding animals are not their enemies. So we live together.' The jeep ride from the airstrip to the lodge will often take tourists through a herd of elephants. 'We still milk our cows,' as one elder on the terrace puts it with a wry smile. 'But now, with the tourists, we can milk the elephants, too.'

The idea of making money from conserving their wild-life began for the Il Ngwesi back in the 1970s, at the suggestion of local landowner Ian Craig, who had befriended the Maasai and later helped them raise the money to build the lodge. Kip was a young warrior then, and initially hostile to the plan. 'I thought it was another way of somebody trying to grab our land,' he says. 'I thought it was colonisation.' He says he was persuaded to change his mind by a Maasai elder from another clan, who told him conservation was the way forward for his people.

Kip is an elder himself now, chairman of Il Ngwesi, and proud to boast of a business model that meets the needs of visitors, wildlife and his people. 'I was brought up in an old culture. I was born in a *boma* [a livestock enclosure] and I don't know how old I am.' Conservation, he says, 'is a way of holding on to our culture while creating opportunities for our people. It is a win-win.' Without it, 'all of us would be in town by now. The poachers would have chased everybody out.'

The key to this harmony, he says, is the community's sense of ownership of their land and the wildlife within it. With ownership comes a sense of responsibility. 'When the community owns the conservancy and the wildlife, it's not going to be easy for a poacher to come through, because people will alert the rangers.' Now Kip spends much of his time advising other Maasai groups on how to achieve what his clan has managed at Il Ngwesi. 'They've seen our success. They think we are professors,' he says with a smile. He wants to link up the different territories 'to make a big conservancy'. That would allow the Maasai to move their cattle more widely, reducing grazing pressure further.

Kip has travelled internationally too, to tell the Il Ngwesi story. 'We have friends all over the world,' he says. He sees himself as part of a global conservation community. 'When I look at conservation, I look at a community of the whole world. We need to work together.'

His vision of how conservation should be done is clear, but it is also a challenge for conventional conservation. His conservation doesn't try to separate people from wildlife, but instead to integrate their lives with the natural landscape. For him, communities have to 'feel ownership of the wildlife' and to appreciate and benefit from it. 'You cannot do wildlife without people. I want to bring nature back to how it was many years ago. With people living in it, benefiting from it, and with wildlife happy and big numbers coming back.'

Kip remembers his grandparents, who taught him the lore of his land. They told him that the clifftop where the lodge now greets tourists was a dangerous place. Lions waited there for their prey. It was a hideout for the shifters, too. 'My grandparents would be astonished to see the lodge there today. But they would be so proud. I always believe they are watching me and guiding me.'

Looking forward, he has a dream: to bring black rhinos back to Il Ngwesi. Even when he was a child, 'they had all been hunted and wiped out.' His grandparents told him about them, but they were as remote to him as stories of dinosaurs, he says. Now 'through our connections with the rest of the conservation community, we are going to bring them back to where they once belonged. It will not be a story any more, it will be a reality.'

• • •

Such stories of nature's restoration are proliferating. Across the world, we are going beyond protecting what remains, towards nurturing its reoccupation of the land. Sometimes this involves sparing land for nature, as in national parks or the bison reserve in Montana. But increasingly it involves sharing the land better with wildlife, as achieved by the Il Ngwesi clan in Kenya. Often, tourism has a big role. But there are many wider reasons why communities and governments seek to protect nature and its ecosystems. Such benefits may range from trophy hunting to preventing floods and droughts, from reducing temperatures to halting soil erosion, and from ensuring there are pollinators for crops to providing pollen for honeybees. Forests are on the front lines of many projects aiming to restore these 'ecosystem services'. Many deforested lands are now in recovery mode.

Half a century ago Costa Rica in Central America was a hot spot of deforestation. Then, President Rafael Calderón called a halt. Inspired by international calls to halt deforestation made at the 1992 Earth Summit in Rio de Janeiro, he began paying landowners to protect and restore their forests. The payments were primarily funded by a tax on polluting fuel. The system, which has been maintained by six subsequent governments, made the country a beacon of environmental restoration. His system of 'payments for ecosystem services' has been copied round the world.

Christiana Figueres, a Costa Rican diplomat who was inspired by Calderón's initiative, became famous as the chair of the successful Paris climate conference in 2015. She sits on The Earthshot Prize Council and we hear more from her later. Of her home country: 'Costa Rica does not have any minerals,' she says. 'Our greatest treasure is our natural resources, which depend on our forest cover.' In Figueres' view, giving thousands of smallholders and ranchers an income for protecting their trees, planting more and giving forests room to regrow naturally has transformed the country.

The revived forests are a magnet for eco-tourists, worth $2 billion a year to the national economy. Former ranchers now build forest lodges amid their new trees. Such is the speed of nature's recovery that visitors are often unaware that, just a decade before, the 'virgin rainforest' they come to see was a cattle ranch. The forests have also prevented floods, protected water supplies in rivers that fill taps and generate hydroelectricity, and contributed to the country's leading role in fighting climate change. Costa Rica is on target to be carbon-neutral by 2050, one of the first developing countries to make that commitment. It is a trailblazer, says Figueres. 'There is no doubt that the world will not be able to address climate change without reforesting a substantial part of the land.'

Costa Rica is far from alone in warming to trees. The old paradigm that clearing them is a requirement of economic development is disappearing. From Panama to Nepal, El Salvador to Thailand, Chile to Vietnam and India to China, countries with fast-growing economies are at the same time more forested than a generation ago. Sometimes there are important strategic reasons for reforesting. In Panama, it helps ensure the Panama Canal, one of the world's most important shipping routes, never runs out of water. Sometimes there are commercial reasons, with new forests created as plantations of fast-growing trees for harvesting, though their benefits for ecology and carbon uptake are generally fairly small.

But some of the most successful stories of reforestation happen when local communities are in charge of their own forests – to both harvest and protect them. In the Nepalese foothills of the Himalayas, more than 20,000 local forest user groups licensed by the government have increased national forest cover by a fifth. 'Twenty-five years ago, the trees here were so sparse you could see people walking on the distant hills,' says Ramhari Chaulagai, the chair of one forest group south-west of Kathmandu. With no trees, the hills were dry and 'our people had to go a long way to get water'. Now they have wood for fuel and to make handicrafts, they can graze animals in the shade of the trees, their wells are full, and they can look out for passing wildlife that includes pangolins, deer and the occasional leopard.

Governments worldwide have pledged to reforest an area bigger than India by 2030. That is 2 per cent of the world's land surface. It is unclear how much of that will be natural forest, or even native trees. But there is a growing movement for rewilding the land that extends beyond conservationists to governments, landowners who want to give back to nature, and entrepreneurs keen to spend their spare cash on nature. Almost a fifth of South Africa is made up of private wildlife reserves, owned by Oppenheimers, Gettys and Gulf investors. Richard Branson has been buying there too. The North Face fashion empire, along with Benetton and Ted Turner, have paid for private conservation and nature restoration in the Patagonian wilderness of Chile and Argentina.

What is clear is that while restoration on the scale needed requires money, often from the wealthiest people and governments, this cannot be a Western plan. As we learned in Il Ngwesi and Costa Rica, true long-term restoration only works when the people that live there are in agreement, in charge, and directly see the benefits. Where people are part of nature's restoration, not separated from it.

In Scotland, Paul Lister, the heir to a flat-pack furniture fortune, is turning a Highlands hunting estate into a 9,000-hectare wilderness reserve, full of the natural pine trees that once covered much of the country. He hopes the stags will one day be hunted by reintroduced lynx and wolves, setting in train a cascade of ecosystem recovery, much as the grey wolves have done in Yellowstone.

As Sir David wrote about earlier, on the Knepp Estate in southern England, fields are being returned to the wild, thanks to the well-named Isabella Tree, who has let nature rip on her 1,400-hectare former farm and wrote a best-selling book about it, *Wilding*. She didn't plant a thing, she says. She just let the thorns create habitat for birds that brought in seeds from which trees soon grew. She holds out little hope of bringing wolves back to a retreat that is just 6 kilometres from the town of Horsham. But beavers are building their dams again on the river that crosses the estate.

'Rewilding has become a trendy buzzword,' says Sophie Wynne-Jones of Bangor University. It can mean different things to different people. Some use it to describe planting a few new trees. But that is wrong, she says. It should mean 'letting nature become more self-willed . . . allowing wildlife the freedom to flourish and habitats to regenerate naturally.' When we embark on rewilding, we should not try to dictate the outcome. It is fine to reintroduce a few lost species to get things going – wolves maybe, but also perhaps fungi, or plants as food sources. But after that, we should let things go, let nature take its course, and expect the unexpected.

We need to rewild our heads, too. Rather than preserving it aloof from us, we need to dive into nature. To recognise that we are part of it. We should embrace it by visiting, by camping in it and eating its fruits. So bring on the school trip to the wild. Move classes into the woods. It means something else too. Rewilding should mean scientists and park authorities stepping back. And remember, Wynne-Jones says, that 'rewilding isn't about rewinding the clock, it's about looking to the future and the challenges nature will face.' It is about reconnecting the fragments of natural landscapes to remake something bigger, where nature has room to breathe, where if the climate does change, species can move with it.

· · ·

To achieve widespread rewilding requires giving up land on a large scale as well as utilising or restoring the huge swathes that have been abandoned. In a world of approaching 8 billion people, that means using the land we continue to occupy much better. The most obvious way to achieve that is by reducing how much we use for agriculture, whether for growing crops or raising livestock. Luckily there is huge potential to do this, and in places it is already happening. Some researchers believe we may already have reached 'peak farmland'. Yes, we are still clearing forests and draining wetlands for farming, especially in the tropics. But elsewhere we may be giving up even more land, where nature may recover its footing – albeit never with the diversity that we are losing in the tropics.

All across Europe, the land area under agriculture is reducing. Often it is simply abandoned as young people leave rural areas for jobs in cities. In the past three decades the countries of the European Union have almost accidentally given up an area of farmland the size of Switzerland. In the Carpathian Mountains from Poland through Romania to Ukraine, 16 per cent of farmland was abandoned in the 1990s, after the collapse of Soviet collective farms. In Russia, the plough has disappeared from an area twice the size of Spain. Irina Kurganova, a geographer at the Russian Academy of Sciences, calls it 'the most widespread and abrupt land-use change in the twentieth century in the northern hemisphere'.

When farmers leave, nature usually recolonises with trees. Nature's recovery on abandoned countryside is the major reason why Italy today has a million hectares more trees than a generation ago, and why ibex, wolves, vultures and wild boar are returning in northern Portugal.

As in North America, wolves have become the symbol of returning nature in Europe. Once hunted down ruthlessly and corralled into old refuges in Eastern Europe, they are spreading west. Individual wolves have been tagged and tracked on journeys through the Alps. Populations are growing in France, the Netherlands, Germany, Italy, Spain and Portugal. Western Europe may now be home to more than 12,000 wolves. Joining them are lynx spreading from the Carpathian Mountains, as well as jackals, brown bears, wolverines, ibex and even the European bison, and also hundreds of thousands of beavers and probably millions of wild boar.

Most of this resurgent wildlife is not living in large pristine landscapes. The animals are creeping back on to their old terrain along railway tracks, through pockets of woodland, over farms and abandoned industrial sites, and hiding in suburban gardens. They are nature on the march in one of the most densely populated parts of the planet.

But before Europe congratulates itself too much, we should remember that this great restoration arises in part because Europeans buy a lot of food from abroad, including from tropical countries that are still being deforested. By some estimates, growing crops to feed Europe is responsible for more than a third of tropical deforestation.

So for Europeans and North Americans to genuinely give land back to nature – and to allow the rest of the tropics to join pioneers in reforestation such as Costa Rica – more must happen. In particular, it requires reducing the amount of land used to feed the rich world, wherever that land may be. So how can we reduce this human foot-print? This is a complex question requiring big changes in what we eat, how we produce what we eat, and also whether we can share the land that we need to grow and raise our food better with nature, as well as spare more land for nature.

• • •

An obvious revolution is needed to reduce food waste. A third of the food the world's farmers grow never gets eaten. It is left in fields, rots in warehouses, is lost during shipping, is discarded by supermarkets for being the wrong shape or beyond its sell-by date, or is simply scraped off our own plates into the bin. If we ate up, the world is already producing enough food to feed 10 or 12 billion people on their current diets, albeit at immense environmental cost.

Many of us also need to change our diets, in particular to rely less on livestock. This is a huge issue. Switching the world to a plant-based diet, without animals as protein intermediaries, could at a stroke reduce the global requirement for farmland by as much as three-quarters. It would also massively cut the amount of water taken from rivers to irrigate crops. A litre of milk typically requires between 2,000 and 4,000 litres of water to grow the crops to feed the cows. David Attenborough calls beef the planet's single biggest cause of habitat destruction. The Amazon rainforest is being cleared so bare ground can grow grass for raising a meagre one head of cattle per hectare, or to grow soy to feed to animals in European factory farms. That is madness.

There are, of course, examples where the production of meat is an integral part of nature's restoration – where domesticated breeds of pigs or cows replace and mimic the herds of wild grazing animals that have been largely wiped out in Europe and North America. But we need to realise that sustainable production cannot serve anything like the scale of current Western consumption. Unless radically different methods of producing meat are developed, meat is likely to have to become a rare part of most of our diets, if nature is to return.

We will return in Part 3 to look in more detail at how we might reduce our personal land and water footprints by changing our diets. But this is not all down to our personal choices. Technology can also help to reduce our footprints, by altering how the food we buy is produced.

The twentieth century saw a revolution in how the world grew its food. That revolution was driven not by fear of ecological breakdown but by the fear of global famine. This was a time when the world's population was doubling every generation, and nobody knew how they would all be fed. In the 1950s and 1960s scientists came up with a 'green revolution', developing high-yielding varieties of critical food crops such as wheat, maize and rice, and these strains are now grown throughout the world.

This revolution doubled, and in some places quadrupled, the productivity of existing farmland. It spared land. Without those new crop varieties, we would have a lot less room today. But this intensification of production has had a major downside. Almost all the high-yielding crops of the green revolution require very high inputs of fertiliser, pesticides and water. So while they have saved land and kept the world fed, they have emptied rivers, polluted the environment with pesticides and nitrogen fertiliser, and consumed large amounts of energy in making that fertiliser. These high demands were a design flaw in the new crop varieties. Tonnes of food per hectare was all that counted.

Nobody back then was worrying too much about the environment. But we are now. Modern crop breeders say there is great potential to produce high yields with much lower inputs, and with much less pollution.

Whether we can square this circle may become clear soonest in the Netherlands, a hotbed for agricultural technologies. The Dutch have a reason to worry about where their food comes from. Thousands starved to death there during the Second World War. Since then, they have pioneered new techniques to secure their food supplies. As a result, despite being one of the smallest and most densely populated countries in Europe, the Netherlands has largely ended food imports and become one of the world's largest exporters of vegetables. Its high-tech farms often still have high chemical demands and environmental impacts. But some pioneering Dutch farmers are now rethinking things.

Precision farming is the name of the game. The aim is to reduce inputs and cut pollution by being much smarter at meeting the needs of each plant. So drones fly over wheat fields, detecting the ultraviolet light coming from each plant. It is a way of spotting unhealthy leaves, showing they may need extra water or fertiliser. That way the farmer can target what each plant needs, rather than pouring on more just in case. Robots shuffle down rows of crops delivering granules of fertiliser or water to plant roots just when and where the farm computers says they are needed. Other robots identify and zap weeds without recourse to pesticides. The robots take only a tenth of the energy required by big-farming kit, and do the job with much greater precision and efficiency.

Some believe it may soon be time to get rid of conventional fields altogether. Greenhouses for growing fruit and vegetables are nothing new. But recently they have been dispensing with soil and suspending plant roots in liquid growing medium. They often dispense with the sun too, in favour of 24-hour immersion in LED lights. With no soil or sunlight needed, you can build vertical farms to cut land needs further.

A vertical farm near Copenhagen in Denmark will produce a thousand tonnes of salad vegetables a year, a fifth of national consumption, on an area no bigger than twenty soccer pitches. Its Taiwanese tech grows crops in shallow troughs of nutrient-rich water stacked fourteen layers high. Tiny robots deliver seeds to the stacking shelves; sensors monitor the environment and control the lighting. There are fifteen harvests a year. The vertical farm uses less than 1 per cent of the amount of water needed to irrigate in fields. It does need lots of electricity, especially for the lights that drive the growth, but perhaps no more than would have been used to manufacture the fertiliser applied on regular fields. In any case, its energy comes from offshore wind turbines. So the carbon footprint is minimal.

Another version of the same idea comes from Vertical Future, a start-up based in the unlikely agricultural suburb of Deptford in south London. This 'farm' is stacked high with shipping containers that founders Jamie and Marie

Burrows have turned into miniature high-tech farms. They come complete with LED lighting, climate control, irrigation, growing medium – and all the software needed to run them. Just add seeds. Each container can grow 5 tonnes of fresh produce in a year. These pop-up farms can be set up to deliver food wherever and whenever there is demand – inside a supermarket, maybe.

Similarly fieldless is Keiran Whitaker's basement farm-cum-factory beside railway tracks behind London's Tower Bridge. His basement is full of insects gobbling up organic waste supplied by local craft breweries, rotting food from supermarkets, and coffee grounds from city cafes and bars. They grow into fat larvae that provide protein to feed to farm animals. Keiran's factory flies replace fodder crops from fields and fishmeal from the oceans, sparing land for nature and fish from the sea. He is ambitious. 'Our mission is to restore the natural world by revolutionising the food supply chain, feeding animals on insects,' he says.

This is not just talk. Keiran is passionate about the environment. As a young man, he travelled the world teaching scuba diving. He looked around him. 'I've seen how coral reefs are dying, how oceans are fished to extinction, and how our rainforests are being chopped down everywhere,' he says. He saw how much of the environmental destruction was caused to feed us. 'I couldn't take any more,' he says. 'I wanted to do something that would fundamentally change the way we feed the planet.'

His inspiration came from his old life back home, where he remembers his mother always composting the family's food waste to grow vegetables in their garden in Lewisham, south London. Now his insect-growing company, Entocycle, does the same thing, recycling food waste to make more food. 'We just need to replicate the old ways,' he says. 'But now we need to do it at a truly massive scale.'

Keiran chose for his great task black soldier flies. They are common around the world, well known for feeding on compost heaps and waste ground. They are among the most efficient converters of organic waste into insect protein. In Keiran's hands, they are on the march. He breeds them under controlled conditions, keeping 5 per cent to produce the next generation while feeding the rest on the food waste from hipster London. 'In ten to twelve days, they grow from the size of a grain of sand into an inch-long purple protein bar,' he says. 'They will eat almost any organic waste. We are showing it is not waste at all, just an unutilised product. We are closing the loop.'

Insect farming 'has an incredibly low environmental footprint', Keiran says. But in his hands, it is undeniably high tech. His factory has a tightly controlled environment, with lighting, temperature and density of insects all subject to the dictates of his entomologists, who supervise round the clock. 'You have to be able to control millions and millions of insects through their life cycle. But it means you can turn food waste into sustainable protein anywhere in the world.'

His fly larvae are made into pet food, but the big potential market is the 70 billion animals that we turn every year into food for feeding ourselves. Most people like at least some of their protein in the form of meat. So Keiran's technique and others like it offer massive potential to stop feeding livestock from crops grown in fields, and to give that land back to nature. The WWF says insect protein fed to British livestock alone could allow the rewilding of hundreds of thousands of hectares of land currently used to grow soy in Brazil.

Other innovators are asking why we need those 70 billion farm animals at all. Why not make 'meat' in the lab? Artificial meat made from plants is improving in quality, but doesn't generally have either the taste or texture that carnivores crave. It still only makes up 2 per cent of the global meat market. Some technologists believe they may do better by avoiding the plant route and growing what they call 'cultivated meat' from animal cells. This is still pioneer food technology, but it may be closer than we think. Shulamit Levenberg of the Israeli Institute of Technology says she is close to cooking up what she calls 'bio-print' rib-eye steaks, with the texture and taste of the fat and muscle of real steak.

The idea began as an offshoot of growing human tissue for medical grafts and transplants. 'One of my students asked if we could try to grow tissue from cattle,' Levenberg says. 'We knew that industrial-scale livestock farming was a huge driver of the climate and biodiversity crisis, so we thought if we could culture beef, it might have far-reaching positive impacts. We gave it a go, and Aleph Farms was born.' In Jewish lore, the letter aleph represents the oneness of God.

Her 'farm' takes cells from a healthy cow, and grows them in the lab. 'They mature into different cells, like the cells that comprise the muscles and connective tissue of cows,' she says. 'Then we have meat, similar to a steak grown from a cow grazing in the field. This cellular agriculture doesn't use antibiotics and we don't kill a single cow. It only takes three weeks, compared to two years growing meat conventionally.'

Others are working in the same field. But Levenberg hopes the key to delivering the best-quality cultured meat will be her unique 3D bio-printer, which assembles the animal cells into meat with the textural feel of a steak. 'It's a way to be able to continue to eat meat, but drastically reduce harm to the planet.' The work is still experimental and expensive. But scaling up will reduce costs drastically, and she plans to launch her first product, a thin-cut beef steak, in 2022. Then a rib-eye steak will follow. 'I believe within just a few years we're going to see cultivated meat in supermarkets and restaurants everywhere.' If so, we may see yet more fields being abandoned.

Can we imagine a world without fields? The most futuristic – and arguably greenest – approach of all to making food without fields may be under trial in Finland, where they are dispensing with animals and even plants to make our sustenance directly from renewable energy and air. The trick, says Pasi Vainikka, the founder and CEO of a start-up called Solar Foods, is hydrogen.

Vainikka began as an energy researcher. At Solar Foods, he uses electricity to split water into hydrogen and oxygen. The hydrogen is then fed to bacteria growing in tanks containing water dosed with carbon dioxide extracted from the air and some nutrients and vitamins. The bacteria grow and multiply, much like Keiran's black soldier flies, creating a mass of protein-rich organic powder similar to dried soy or algae that he calls Solein. This, he says, can be included as a protein-rich ingredient in processed foods ranging from bread and pasta to lab-grown meat.

The basic process is quite similar to photosynthesis, the means by which plants grow, but it is much more efficient. The protein product could in future make lab-cultured meat, or produce vegetable oils that replace palm oil, says British environmentalist George Monbiot, a fan. The land gain from this field-free food production is staggering. 'Cultivating all the protein the world now eats with their technique would require an area [of land] smaller than Ohio,' Monbiot says.

Admittedly, the process requires a lot of electricity. Solar Foods gets this from solar panels. They require a lot of land, but only a tenth as much as would be needed to grow the food in a conventional fashion, says Vainikka. His method of feeding us is, he says, ten times more climate friendly than most plant-based proteins and about a hundred times more than meat. Moreover, it is the first food produced without either farms or fossil fuels. By simultaneously cutting out our two biggest footprints on the planet, it could, Monbiot says, 'change our entire relationship with the natural world'.

• • •

That may be a somewhat utopian vision. Maybe for the twenty-second century. We are not all going to be eating solar food anytime soon. So while better food technologies

125

have great potential to spare land for nature, we will still need to find immediate ways to share the land with nature much better than we currently manage. Thanks to nature's great capacity to adapt, there is more potential to do this than you might imagine. Even on intensely used land, nature can get by and even thrive. Many birds love farmland, for instance, because it provides seeds to eat. In the Amazon, tapirs and armadillos leave the rainforest to forage for food in nearby fields. If farmers can connect up remaining patches of wilderness, this too can help nature survive and even recover.

So here is an unlikely hero. Mislin Elahan is a palm-oil farmer in Sabah on the tropical island of Borneo. Once a byword for untamed jungle, Borneo has lost almost half its forests in the past half-century, mostly to the ubiquitous vegetable oil that Mislin grows. Many people see farmers like her as environmental villains. But that's not how she sees herself. Most days, she is out in the fields with her family, collecting bunches of the fruit on land that was once rich rainforest. But often she puts binoculars round her neck, waves goodbye to her children, and heads off on her moped to the nearby forest, to check out the orangutans. She has been studying them for twenty years, ever since the forest here was converted to palm-oil plantation.

Mislin works for Hutan, a conservation project set up back then to study the animals and their changing habitat. She has co-authored academic papers showing that, if established and managed correctly, orangutans can live and move around in the palm-oil plantations. They build nests in oil palm trees, and eat the fruit and young shoots. About 800 remain, of the 4,000 who once lived locally. But they need access to natural forest, and generally stay within 50 metres of it. Even so, Mislin concluded in one study that if some wild forests remain nearby, 'orangutans appear more resilient to habitat changes than was originally thought'.

We should not write off landscapes like hers as death zones for wildlife. While the large scale plantations themselves are hostile to most wildlife, damage can be reduced with well-planned restoration. She says her purpose now is to persuade her fellow farmers to restore patches of forest that reconnect the surviving fragments, creating natural corridors for the animals to move around and meet other groups. She says if the rest of the world truly wants to save Borneo's orangutans, we should stop seeing people like her as enemies and start seeing them as potential collaborators in conservation. As a farmer herself, Mislin proves the situation can improve if we maintain farms alongside nature.

Increasingly today, the world's surviving orangutans – as well as many other forest creatures from tigers to elephants – survive in areas where there are either timber or oil palm plantations. It's not ideal. Far from it. But it is becoming clear that many species can survive in such conditions, and even adapt to them. So it makes sense to find ways of sharing the land with them, says British orangutan conservationist Cat Barton of Chester Zoo, who has collaborated with Mislin and her colleagues for years. 'In Borneo you can drive for hours though oil palm plantations,' she says, but wherever you find a patch of forest you 'get orangutans, proboscis monkeys and hornbills, all living naturally'. They may appear to be doomed populations, but they could be the start of nature's recovery.

Barton, who comes from a European country that was itself once forested and home to bears and wolves, says it would be hypocritical of the West just to try to ban Borneo's palm oil. In any case, 'palm oil is here to stay. It is a wonder crop. It is in such a wide range of products – biscuits to soaps, candles, pet food and cosmetics – that getting rid of it would be virtually impossible.' And it has much higher yields than most of its rival vegetable oils, such as sunflower or corn oil, producing more than 5 times the yield on the same land. 'So if we switched to other vegetable oils, we would need far more land to grow the same amount of oil,' she says. The goal of land sparing suggests we should reduce unnecessary waste and consumption, but for the things we do need, stick with palm oil, while making its production greener.

Barton remembers that when she first met Mislin,

we were in absolute awe of her knowledge of the rainforest. She knew every species, every bird call, and was so dedicated to her work. She showed me that palm oil is not a bad crop, it's the way that it is grown that can be bad. Nature can fix itself, but we need to give it that little push.

Is Mislin's world a metaphor for humanity's future on the planet? We need resources from the land. We should get them by farming efficiently and intensively when we need, and creating room for nature to thrive elsewhere. Sparing and sharing sound like opposites. In some ways they are.

But as Mislin shows, intensifying crop growing on one field can allow a farmer to share more of the rest of their land with nature.

Sharing agricultural land with nature often goes under the title of 'agroforestry'. Arguably, Mislin is an agroforester. Agroforestry involves either putting trees back on farms – to provide shelter and cooling, pollinators and pest control, wood and other commercial products, and migration corridors for wildlife – or growing crops in the shade and cool of existing forests. Trees can sustain crops and be crops in their own right. Either way, agroforestry benefits both farmers and nature.

Agroforestry is not new. Ancient methods of shifting cultivation in forests have always been good for nature as well as for farmers. Many other traditional farming systems grow bush crops such as coffee, vanilla and cacao (which makes chocolate) in gardens that rely on forest shade. Researchers have found that these systems sometimes have more biodiversity than the surrounding forests, and even when they don't they connect up natural forest areas.

Some forms of agroforestry have gained the label 'farmer-managed natural regeneration'. This originated in the arid Sahel region of Niger in West Africa, where farmers in the 1980s began nurturing trees that grew naturally on their land. This was contrary to official government advice to farmers to dig up tree roots, because they made ploughing harder. The Niger farmers found the trees benefited their fragile arid soils, provided additional crops of wood and leaves, and gave livestock and people alike shade and shelter from the sun and wind.

It happened by chance, when two young farmers in Dan Saga, a village in the remote Maradi province, returned late from jobs in distant mines. The rainy season had already begun, so they planted crops without first clearing the ground, like the rest of the village had done. To everyone's surprise, their harvests of maize, sorghum and millet turned out better than their neighbours'. Other villagers copied them. Soon the secret was out and the idea took off. Today, natural trees grow on some 6 million hectares of fields through Niger and into Mali, and satellite images show areas that were once yellow and probably turning to desert have changed to bright green.

Now, everyone wants a slice of the action. Governments in Africa see agroforestry in general, and 'farmer-managed natural regeneration' in particular, as the best way to hold back deserts, fight soil erosion and sustain land that is badly needed by their fast-growing populations for growing food. The African Forest Landscape Restoration Initiative (AFR100) plans to create 100 million hectares of agroforests in Africa by 2030. 'Agroforestry has the potential to increase food security for 1.3 billion people,' according to the UN's plans for a decade of ecological restoration. In future, some say, Africa may have fewer forests but more trees.

· · ·

The story of the farmers of Niger shows that the revival of traditional methods of farming can often help both people and nature thrive. Many rural communities have ways of traditionally managing their lands that are very different from conventional twentieth-century Western ways. Their methods may appear low-tech, but often achieve high yields through harnessing rather than fighting nature. Increasingly, Western agriculturalists want to learn from them about how to farm among trees or in wetlands, how to conserve soils without artificial fertilisers, and which varieties of crops work best in particular localities. The great international seed banks that half a century ago bred the high-yielding crops of the 'green revolution' are now searching through their stores of ancient seeds culti-vated by farmers over the past few centuries to find ones with long-forgotten attributes that could be vital in a world of climate change.

But at the same time as this reassessment of the old ways of farming is going on, the people with these secrets are losing their land, and are forgetting the wisdom in their traditional techniques. This will be bad for conservation as well as for farming.

It is often said that indigenous peoples and some local rural communities are better at conservation of their own land than outside experts – because they know it better, cherish it more and have superior skills in managing it for people as well as for nature. They must be doing something right, because around a fifth of the world's most vital areas for biodiversity are under local community control. And many more rural lands are in effect managed by them.

From the pygmy peoples of the Congo to the Inuit in northern Canada; from the Wampis of the Peruvian Amazon to the Karen of Myanmar, who created their own Peace Park; from the Zhuang people of southern China to the Manobo island dwellers of the Philippines: recent research shows that the lands of indigenous peoples generally have more species, and their ecosystems are better protected, than those under state management.

This revelation is pushing many conservationists to join calls for such communities to both have their rights better recognised and respected and be given greater control of their lands and the resources they contain. Indigenous reserves, they say, often do a better job of conservation than national parks. By being emptied of people, national parks often become sitting ducks for takeover by loggers, poachers, miners and cattle ranchers. But under local control, they will usually be better protected. So while the world's governments have a goal of designating 30 per cent of the land as protected areas by 2030, it may be even more important to protect and enlarge the control of indigenous peoples and other local communities with strong connections to their land. Here is one example of why this is so important.

The ancient forests of Petén, in northern Guatemala, are home to jaguars, pumas, eagles and crocodiles. Here, thirty years ago, the government created the giant Maya Biosphere Reserve. Within the reserve, it set aside strictly protected forests. But it also created eleven areas for local people – descendants of the ancient Mayan civilisation – so they could pursue their livelihoods in 'community forests'. The division of the reserve was seen as a compromise between the needs of people and the needs of nature. The idea was to keep people and nature as far apart as possible.

The result has been both more alarming and more interesting. Today, much of the reserve has been invaded by ranchers and drug traffickers. The strictly protected areas have suffered badly. Many of them are now cleared of trees, replaced by cattle ranches set up by drug barons keen to cover their tracks as they take their produce across the border into Mexico. Wildlife guards were no match for them and their henchmen. But remarkably, the Mayan community forests remain almost entirely intact. The Mayans defend their land fiercely and the ranchers and drug gangs stay clear.

The Mayans make good use of their trees. They cut timber to make furniture that they sell in a local store. They harvest and sell nuts from trees planted by their ancestors a millennium ago. They provide mahogany to guitar makers in the United States. But there is little indiscriminate cutting. They have management plans for the forests that preserve the trees for the use of future generations. They have fire crews preventing the blazes that often occur outside their land. 'The forest is an economic asset to the people,' explains Juan Girón, deputy director of the Association of Forest Communities of Petén. 'If the person benefits from natural resources, he or she sees them as an asset . . . leading to better care of the resources.'

They are far from alone. In the Amazon too, if you compare a map of indigenous territories with satellite images of forests, the overlap is clear. Indigenous areas are forested; the rest is disappearing.

While many communities draw on their ancient traditions to manage and benefit from their forests and other wild

133

lands, they increasingly draw on modern technology to document their land, police their boundaries and report invasions. The Wapichan people live in a remote region of southern Guyana the size of Wales that borders on the Amazon. They have been engaged in a long exercise to document their land and how they use it, as part of their case to government to be given full title to their traditional territories. 'In the early days there was no GPS,' says former chief Tony James. 'We just walked, following government maps and adding detail to them from our knowledge. But with GPS we have been able to do things much more precisely.'

Angelbert Johnny of Shawaraworo village was one of the digital mappers. It took years to fully document everything, including the wildlife, he says. 'We assembled people in each village and found the elders and experts on the creeks, forests and mountains. They were our guides as we walked, or took boats and bicycles and horses to survey the land.' It was hard, he says. Often they had to put their phones on poles and push them up through the forest canopy to get a signal. 'People got bitten by snakes. One guy had to be dragged out of the bush and given bush medicine to cure him. Another got lost for two days. But we went everywhere and mapped everything.'

They called it the 10c project, after an article in the UN Convention on Biological Diversity that says governments should 'protect and encourage customary use of biological resources in accordance with traditional cultural practices that are compatible with conservation and sustainable uses practices'. Currently only a fifth of their territory is formally recognised by the government in Georgetown.

So the mapping is about land politics. But it is also about day-to-day conservation. The Wapichan rangers return regularly to the bush to report invasions of their land. They can pinpoint to within a few metres the location of illegal gold mines or places where cattle rustlers and others cross their southern border from Brazil. The patrols have a deterrent effect, says James. 'The rustlers fear what they call the "monitors with smartphones". They turn back if they hear we are around.'

GPS is also helping map the rich wildlife of their unfenced woodlands, savanna grasslands and wetlands. Conservationists say the area resembles Africa in its woody wildness, with jaguars hunting for deer. The Wapichan want to set up camera traps to monitor jaguars, in the hope of creating a jaguar reserve. Their land has rich birdlife too, including harpy eagles, pearl kites, savanna hawks and a recently discovered population of a small bright orange finch called red siskin. For a while thought extinct, the red siskin were discovered recently in larger numbers by Wapichan rangers working for the South Rupununi Conservation Society, a body set up by their leaders to bring in wildlife tourists. This is indigenous conservation and environmental restoration in action. There is not a government ranger within a hundred kilometres.

• • •

Most of this chapter is about land and the nature that lives on it. But we face similar choices about sparing or sharing the water that flows across the land. Our rivers and the wetlands they nourish are great ecosystems in their own right, and among the most damaged of all by human activities. Rivers across the world suffer pollution from industry, sewage and increasingly from run-off from farms. There are vanishingly few unbarricaded rivers anywhere in the world. In Europe there is one dam, barrage, weir or other barrier on almost every kilometre of river. Migrating fish are frequently left stranded. The loss of wildlife from freshwater ecosystems over the past century is greater than for any other type. Some of the world's greatest rivers no longer reach the ocean because we take every last drop. Most of the world's wetlands have been drained or cut off from their rivers.

To restore nature to our continental land masses, we need to rewild our rivers too. The task is harder because rivers flow, from the mountains to the sea, often over thousands of kilometres, connecting up small streams, floodplains and wetlands that drain vast areas. You can fence off a piece of land to protect what it contains or to allow ecosystems to recover. But a river system cannot be fenced. You cannot protect a river without protecting everything upstream.

We are careless with our rivers. Too often, dams are left in place even when they serve no purpose, because it is expensive to tear them down. Wetlands are seen as wastelands rather than as being among the world's richest ecosystems. But there is a growing movement to restore rivers and tear down dams. Often the fisheries that return have greater value than the hydroelectricity or irrigation water they once supplied. In France, dams on the Loire, such as the Maison-Rouge, which had blocked salmon and eel migrations, are now gone. Engineers are recreating what environmentalists long demanded – a 'Loire Vivante' or a living Loire.

The European Union Water Framework Directive requires all rivers to be returned to a 'good status'. Though the phrase has never been defined, it clearly means more than just ending pollution. One of Spain's largest rivers, the Duero, is being cleared of dams. Britain has a National River Restoration Inventory, and has promised to rewild 1,500 kilometres of rivers. That will require not just a clean-up and a removal of barriers but also recreating old back channels and meanders, revegetating banks and removing levees that cut rivers off from their natural flood-plains.

Even where we cannot tear down dams, we can reduce their harmful effects with bypass channels for migrating fish, and by releasing some water when nature needs it to save fish and mimic the seasonal river flows that maintain wetlands.

As on land, rivers and wetlands are also often best defended by their local inhabitants. For thousands of years the Marsh Arabs of southern Iraq protected the Mesopotamian marshes from outsiders. Saddam Hussein sought to expel them by draining the marshes, but after his demise they have returned and remain fierce opponents of future drainage plans. Exiled marsh inhabitant Azzam Alwash, an engineer, has worked to bring the marshes back to their former glory by restoring flows down the Tigris and Euphrates rivers, which feed it. Meanwhile, in the

Pantanal wetland of southern Brazil, native Kadiwéu people hold back the tide of ranchers and dammers. The Meitei people in the Indian state of Manipur battle government police and hydro dam engineers to keep control of the magical Loktak Lake, with its floating islands of grasses that are central to its unique ecosystem.

Achieving all this will require a much more rational use of water. As with land, the biggest barrier to rewilding rivers is agriculture. Most of the water we take from rivers goes not for industry or even to fill our taps, but to irrigate crops. Around two-thirds of our abstractions are channelled down canals and diverted on to fields, where some at least gets to the roots of crops. But the wastage is huge. Most of the water evaporates into the air or flows away underground.

The world can do massively better. The days when shortages of water were always met by bigger engineering projects to supply more should be over. The need now is to manage demand, not supply. And to recycle water where we can. Spraying water is usually better than flooding fields. Drip irrigation, which delivers water from pipes close to plant roots, is better still, delivering improvements of 60 to 80 per cent. And smart farming, in which computers assess crop needs and robots deliver their inputs, can do better still.

Even for farmers without robots and computers, there can be smarter ways. Pilot studies show that satellites can keep a watchful eye on fields, too, alerting poor farmers by text message if they are over- or under-watering their crops. Hundreds of millions of cubic metres of water could be saved in this way each year in parched regions of India and Pakistan alone, says researcher Faisal Hossain of the University of Washington, in a recent academic paper.

Meanwhile, water that is lost from soils to underground can be pumped up again. And we can often use sewage from cities, suitably cleaned up, to irrigate crops. In Israel, long a leader in farming in deserts, 70 per cent of irrigation water is now recycled sewage. The sewage contains valuable nutrients, too, a topic we will return to later.

• • •

Here is a final place you probably haven't thought about as a key to resuming a better balance in our relationship with nature. Cities are both hotbeds of agriculture and magnets for some wildlife.

One of the untold success stories of modern farming is how much food is grown in urban areas. Although notoriously hard to measure, by some estimates, a fifth of the world's food comes from farms within city catchments. Whether these are set up on abandoned land or roadside verges, former sewage works or back gardens, parks or allotments or in larger commercial operations, in some cities, farming is the biggest industry. In Shanghai, almost a million residents still work the land. Hong Kong produces two-thirds of its own poultry. In Britain, a million people tend urban allotments. The UN Development Programme reckons urban vegetables, by being grown intensively, use less than a fifth as much irrigation water and a sixth as much land as conventional cultivation. They are sparing water as well as land.

Cities are surprisingly full of nature too. Street trees and parks provide a canopy covering over 10 to 20 per cent of many urban areas. London has a higher tree cover than several English national parks. Cities can be richer in birdlife than agricultural areas. Rather than being dominated by single crops and suffused with pesticides, they provide many habitats and food sources. Rare insects and plants love their unexpected nooks and crannies, including landfills and old industrial sites, as well as parks and gardens. Barn owls and bumblebees, orchids and newts, foxes and falcons, raccoons and wild boar: all enjoy city life.

In North America, many species of mammals are more abundant when humans are around. Footage from thousands of camera traps has revealed that almost twice as many species like us as loathe us, including wolves and black bears, wolverines and bobcats, foxes and coyotes, pumas and elk. Often our occupation of the land provides food or other useful stuff. Justin Suraci of Conservation Science Partners concludes: 'Small, less carnivorous and faster-reproducing species tended to do better with increased [human] activity'.

We should do more to encourage all this. Some cities already are. Utrecht is restoring its canals to create green space; Leipzig is turning old open-cast mines into networks of lakes; Malmö is creating a network of green roofs to hold back storm rainfall and prevent flooding at ground level; Barcelona is cleaning up its rivers; Melbourne has created a database of every tree in the city so people can send emails to them; fish have returned to the cleaned-up River Thames in London.

Singapore is known as a high-tech metropolis. But urban planning visionary Esther An is trying to create a greener urban landscape across the island state, replacing walls, streets and neon with trees and meandering rivers that cool the air, secure water supplies, calm city workers and bring back wildlife. Public housing projects now have community gardens aloft. The city has 3 million trees, including a recreated rainforest. Otters recently took up residence on one of her cleaned-up riverbanks. 'People are healthier and happier when they have access to nature,' An says.

Cities are well known as 'heat islands'. Their concrete and asphalt hold on to heat and make summer temperatures several degrees warmer than in the surrounding countryside. But evidence is growing worldwide that bringing nature back to our cities reduces temperatures and cools us in other ways – making us calmer as well as less sweaty. Trees cut urban noise too, and reduce everything from asthma and obesity to depression, anger and anxiety. Many Japanese indulge in *shinrin-yoku*, or forest bathing, as a way of taking the stress out of life. Philadelphia in the United States reckons that if it increases its existing tree cover from 20 to 30 per cent as planned by 2025, it will prevent 400 premature deaths each year from then on.

The distinction between doing good for humans and doing good for nature disappears in cities. What is good for the squirrel, the salmon or the sandpiper is good for humans too. No wonder that hard-nosed property developers are demanding green space in business districts as an inducement to attract the best executive talent, and to prevent those already there from burning out too young.

None of this says we don't need the wild lands and large pristine places. Of course we do. But we shouldn't be too surprised that when we take over nature's places out in the countryside, then nature will fight back by moving into our urban spaces.

Revive Our Oceans

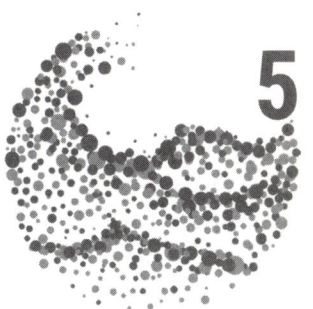

T he ocean feels like home to me. I've always lived
close to the water – I was born in a coastal city,
in the Caribbean part of Colombia, and now my
family and I live in Barcelona, next to the
Mediterranean Sea. So it's always been like an extension of the backyard for me, and now for my children.

After a bad day, I get in the ocean and all my
troubles wash away. I've recently got into surfing, and
when I'm out in the water, waiting for that next wave,
I feel so calm, like time has stopped and I'm at one
with the world. The ocean gives you perspective.
And it should give us perspective on what we're doing
to it.

Our first job is to properly understand what exactly
we are doing to harm it, from pollution to overfishing
to plastics. It would be easy to get depressed about
the state of the ocean, but it is not yet defeated. The
ocean has enormous resilience. We see evidence of
that, even today.

Our hope lies with the brightest minds working on technological solutions, and in teaching our children so that future generations will treat the world differently. My sons, for example, have learned so much in school about ocean life and the planet, and how to conserve and take care of the environment. My oldest son in particular is really engaged with how we can protect the planet – he asks questions all the time, and he inspires me to do everything I can to educate myself and educate them about how we can do our part. Both of them are disgusted when they see plastic on the beach. They often stop and pick it up and dispose of it properly. I don't think I had that kind of awareness as a child, and I'm glad they do.

I feel that we're finally starting to pay our planet the attention it deserves. We now have the tools to communicate with each other instantaneously, and people are voicing their opinions and exchanging ideas. So I feel that we've never been better positioned to turn the tide.

The magic of the ocean is that it can recover. We are in a critical decade, but it's not too late. There are more great whales in the ocean now than there were when I was a child. A lot more. This is because the generations before us had hunted most whales to the edge of extinction. But then the hunting stopped. We allowed the whales to recover. Now we need an even bigger commitment, to end the destruction and degradation of all marine life. I believe this is possible, and what gives me the most hope is all the people around the world who are already taking giant steps forward to make this happen.

With so many inspiring people acting to revive our ocean, it makes me want to do all that I can too. If we all work together, just imagine the future we could create. Imagine if we ensured that we used every part of the ocean sustainably, how rich our seas would be. We'd finally get to see the sea the way it should be. Bursting with life. An enormous wilderness that would provide humankind with plenty of food, if we take care with how we source it. A habitat that would grow and thrive and, in doing so, help us defend ourselves from storms and control the climate. A wonderful blue world that would continue to enrich our lives. The ocean has already given us so much. Now it's time to give back. We still have so much to do, but now is our chance to get it done.

What are we waiting for?

Shakira Mebarak, 2021

The Amazon contains a third of all the tropical rainforest left on Earth and a tenth of the world's biodiversity. It is one of the biggest carbon sinks in the world and home to over 2.5 million indigenous people from an estimated 350 ethnic groups.

Part of a team conducting a lion census in Tsavo East National Park, Kenya. Expanding human populations are encroaching on the lions' former territories, so finding solutions that work for both animals and people is vital.

An Indian Rhinoceros photographed in Nepal. Like almost all species of rhino, they are threatened by human and livestock encroachment and poaching. However, Nepal has seen recent success with conservation efforts and increased their Indian Rhino population.

A mining operation to extract sand in Rondônia, Brazil. Many people are unaware that sand mining is one of the most environmentally damaging practices on the planet.

Aerial view of a road and buildings built around a forest in Hong Kong. City planners are increasingly cottoning on to the fact that it's possible – and beneficial – to build around and with nature, rather than over it.

A grey wolf hunting. Wolves have become the 'face' of nature restoration: some fear them, but where they have returned to their historic ecosystems biodiversity has flourished.

A white-beaked dolphin in Northern Iceland. Species of dolphin can be found all over the world and are important indicator species as to the health of the marine environment.

Seagrass on the north-west coast of Wales. The UK has lost approximately 90 per cent of its seagrass. Seagrass restoration globally is a major priority as it is home to a mass of marine life, and is one of the most effective carbon stores on the planet.

Part of Australia's Great Barrier Reef, the world's largest coral reef system. Coral reefs around the world have suffered from bleaching due to rising sea temperatures and the Great Barrier Reef has struggled in many places to recover. Reefs that have healthy fish populations and low pollution levels are more resilient to bleaching.

Plastics found on the beach in St Helena's, one of the most remote islands in the world.

An off-shore windfarm at Crosby Beach, Liverpool. Off-shore wind is growing fast across the world and advances in technology coupled with demand for clean energy suggest this growth will accelerate in the decade to come.

A mangrove swamp in the Lamu archipelago, Kenya. Mangroves, which have in many places been removed for rice paddies, shrimp farms and other forms of agriculture, are proven to protect against coastal erosion and flooding. They are also nurseries for a wide variety of fish.

Oceans cover two-thirds of our planet. Over a billion people rely on them for food. And for all of us, they regulate the climate and water cycle, and soak up much of our pollution. But unlike the land, wide sweeps of the oceans are owned by nobody and are within no national jurisdiction. Under the UN Law of the Sea Convention, national control generally ceases 200 nautical miles from the shore. Away from the major continents and some groups of islands, the majority of the great oceans are global commons. They are afflicted by a tragedy of the commons: they belong to all of us, but are controlled by no one. Anyone may exploit them – in theory – but currently only wealthy multinational corporations and subsidised national fleets can get to them. And no nation has a direct interest, or legal right, to protect them.

Until modern times that did not matter much. Our shallow seas were abundant with fish and our open oceans were home to vast numbers of migrating giants of the deep. Our pollution was diluted so much that it swiftly became invisible. There was plenty for all. You just had to set your nets to harvest a rich bounty. The oceans seemed too big to be harmed by us.

But that is no longer true. In recent decades our growing numbers and industrialised methods of ransacking nature have taken their toll. We have dramatically overfished ocean waters; poured unprecedented volumes of toxic chemicals, nutrients and plastics into them; and we have even disrupted the ocean's basic chemistry by making it more acidic. Fish stocks today are reduced so much that the catch per unit of effort – a standard industry measure of the economic efficiency of fishing methods – has fallen by 80 per cent since 1950. More boats with bigger nets on ever longer journeys are catching fewer fish.

Despite their vast size, oceans are a finite resource, as prone to damage as our atmosphere. We must treat them that way. We must find ways of managing the oceans sustainably before we exhaust them entirely.

In theory, the world knows this. The UN's Sustainable Development Goals agreed in 2015 pledge to 'conserve and sustainably use' the oceans and their marine resources. And the world has a deadline for achieving this goal: 2030 at the latest. This will require doing three things. First, protecting them from poisoning, whether from direct discharge of poisons or via land or air. Second, stopping their over-exploitation, especially by fishers. And third, protecting the coastal ecosystems that are the primary nurseries for most marine life.

Individual nations have been busy providing some protection for the areas of ocean with critical biodiversity or features such as coral reefs. The aim overall is to have 30 per cent of the oceans inside such areas by 2030. So far, marine protected areas cover less than a tenth of the world's oceans, and many are not well protected and continue to allow fishing. But it is a start. While protected areas on land have a history dating back more than a century, doing something similar at sea is new: they only got going in the 1980s. The Gulf of Cortez initiative that we looked at in Part 1 was among the first. Since then, they have proliferated. More than a quarter of the world's coral reefs – around 400 in all – are now under some sort of protection. One major gain has been to ban oil prospecting from around the largest reef system in the western hemisphere, the 300-kilometre-long Belize Barrier Reef.

These small protected coastal zones are increasingly being complemented by big protected areas in more remote regions. Half of these are around distant islands in the hands of the United States, France and the United Kingdom. They include the waters around the Pitcairn Islands in the South Pacific, a British overseas territory with some of the world's most pristine coral reefs, and the US state the Hawaiian Islands, home to more than 1,500 marine species found nowhere else, including the endangered and endemic Hawaiian monk seal.

But what about beyond, on the high seas that make up two-thirds of the oceans and 45 per cent of the entire planet's surface? Since 2017 the UN has been negotiating a high seas treaty, aimed at protecting marine life outside of national jurisdictions. We shall see what it delivers, but one option may be regional agreements between nations near to large, open and currently unpoliced waters. A trailblazer may be joint action to protect the Sargasso Sea in the North Atlantic. It has no land borders but is a vast stretch of becalmed ocean twice the size of India stretching from west to east between the island of Bermuda and the Azores.

It is windless and notorious as a place where sailors

become stranded in the doldrums. But it is also one of the planet's most unique and valuable ecosystems, named after the brown *Sargassum* seaweed that drifts in its slow currents. Oceans activist Sylvia Earle calls it 'the golden floating rainforest of the ocean'. It is so rich in nutrients that a vast array of marine species migrate there to breed, including tuna, swordfish, eels, turtles, marlin, whales and, most mysteriously, eels from Europe and North America, which travel thousands of miles from their river homes to spawn amid the weed. These maritime doldrums are also a perfect place for plastic waste to accumulate. The Sargasso Sea is one of the great ocean rubbish patches. Still, eleven governments, including several Caribbean nations, the United States and the United Kingdom, signed a pact in 2014 to protect it.

• • •

We saw in the Gulf of Cortez that coastal protected areas, the nurseries of the oceans, can be revived. If rigorously protected from fishing and other threats, nature will regenerate quickly. And these areas soon seed life into adjacent areas of the ocean. But we saw too, on the beach in Liberia, how hard it is to police the open waters. Trawlers move in undetected, and fill their nets before anyone notices, still less acts to halt their activities. It needn't be that way. Remote waters can be policed. But it needs both political will and technological surveillance. Luckily, the technology at least has changed out of all recognition in the past decade.

Sitting at a computer in Harwell Innovation Centre outside Oxford, Emma Seal, a British fisheries analyst, has a map of the world's oceans up on screen. It shows thousands of fishing vessels ploughing through the waters, their identifying data and tracks downloaded from satellite images. She is looking for trouble. Her computers have artificial intelligence to make a first assessment of whether the ships' movements are suspicious, helping her to swiftly establish whether they should be there – fishing legally, under licence from national governments – or whether they are fishing pirates, cruising the oceans to snatch fish from local fishers or no-fish zones, often decimating fish stocks and wrecking marine ecosystems in the process.

Today, Seal is watching dozens of ships off Central America for the government of Costa Rica. Most vessels are local fishers going about their legal activities. But sometimes the waters are invaded by super-trawlers on the prowl without licence, able to catch hundreds of tonnes of fish and disappear into the night. Once they could be confident of going unnoticed, but now a phone call from Harwell to the Costa Rican coastguard can instantly send a patrol boat to sea to intercept them. It doesn't happen every day. But in 2020 Seal's colleagues found five of these giant vessels fishing illegally before applying for a licence to fish.

Worldwide, around a fifth of fish are caught illegally, a figure that rises to more than a quarter off West Africa. Many such ships fly 'flags of convenience', meaning they are registered in countries with minimal legal regulation, including identifying the true owner. The most famous of these is Panama, but others include landlocked countries such as Bolivia and Mongolia.

Seal works for OceanMind, a non-profit organisation that

also has contracts with governments in Bangladesh, Thailand, West Africa and elsewhere. Using real-time satellite data provided by the Canadian and Norwegian governments, it tracks any vessels that come into the 200-mile economic exclusion zones of contracting countries. 'If a vessel is hundreds of miles from land, out on the high seas, they can do what they like if there's nobody to stop them – except when technology is brought into play,' she says.

'From the tracking data, we can assess their likely activity, based on [things like] their route, the speed they are going, if it's shallow or deep water, and what the weather and currents are doing,' says Seal. Slow-moving vessels are likely to be setting nets, for instance.

> We also have access to their licences and the regulations in the area. Overlaying all this information enables us to understand what people are doing, whether they are fishing in areas that they're not supposed to fish, fishing species they're not supposed to fish, capturing more than their quota, or using fishing gear they're not supposed to use.

The system often finds repeat offenders. Two Thai super-trawlers apprehended in Bangladeshi waters after being spotted by OceanMind were already being investigated by Interpol for illegal fishing off Somalia in East Africa.

Costa Rican coastguard captain Henry Centeno Mora says the OceanMind technology has changed his work dramatically. Once, catching the illegal fishers was hit and miss, or depended on inspectors stationed on trawlers for months at a time, keeping tabs on their activities, and reporting back to the authorities on their return to port. They were struggling. But satellite surveillance has changed the odds. 'It enables us to monitor huge areas,' he says. 'When OceanMind detect illegal fishing, we set off for the target immediately, board the vessel, and check their legal status. We often find their satellite device is switched off. So OceanMind is a very important tool. It's like having an eye in the sky twenty-four hours a day. Now there is no escape.'

'Protecting marine species is particularly important for our country,' he says.

> They are something unique to us. Boats coming from other countries often use illegal methods. Shrimp vessels pull up anything on the seabed, including small fish. Others encircle the fish with nets and then use explosives and haul them in, in enormous quantities. You see boats that are almost sinking under the weight. This really harms our small-scale fishers who fish according to the rules.

Right now OceanMind has only a small set of clients. But, says Seal,

> there's no point on the oceans that we can't look at. The impacts we've already had are huge, but if all the coastal states of the world did that, it would have an incredible impact on the health of our oceans. We like to describe ourselves as ocean detectives. We make the unknown known and the invisible visible. Technology has completely transformed the way countries can manage their waters.

• • •

There are other ways to use modern technology to police fish stocks. There are eyes in the waters as well as eyes in the skies. With many governments short of funds to patrol their waters or sign up to OceanMind, those eyes often belong to local fishers. They are the people most at risk from the illegal super-trawlers that take their fish, damage their nets and sometimes take their lives. They know what is happening, and have an incentive to reveal all. But they have been generally powerless to prevent these marine invasions. Or, they have been until now. For a smart new initiative from the UK-based Environmental Justice Foundation is putting power in their phones.

Since 2021 the foundation has been trying out a phone app that allows fishers who spot an illegal boat to take a photo with its name or ID number showing, record the location and upload it to a central database, from where government coastguards can take action. The app was first tested among coastal fishers in Liberia, where foreign super-trawlers are regularly invading their waters.

On land, apps using phone cameras and GPS are giving local communities the ability to document and report invasions of their territories. Now they can do the same at sea. On the beach of the Liberian capital Monrovia, one user called Emmanuel said: 'Last year I just used to look on helplessly and hopelessly as the boats destroyed my nets. Now I got evidence.'

• • •

Not all big fishers in the oceans are the same. Not all are bad. The international fleets with big investment in fishing equipment and with large markets to keep supplying fall into two camps. On one side are the pirates with little concern for the future of fisheries, intent on grabbing as much as they can as quickly as they can, before stocks run out. On the other side are those interested in the long term, in being able to catch fish to supply their customers in years and decades to come. They would rather do this the easy way, with plenty of fish to catch, than the hard way, chasing ever-declining stocks. So they have an interest in conserving stocks. The good news is that, belatedly, those with a long-term interest are gaining ground.

In 2017 nine of the world's biggest fishing companies – the ones who dominate tuna, Peruvian anchovies, Patagonian toothfish and other highly profitable catches – promised to clamp down on illegal fishing. The following year the five biggest fishers of krill agreed a voluntary pact to stop catching these crustaceans, which form vast swarms in the Southern Ocean, in ecologically vulnerable waters off the Antarctic peninsula. This is reputedly the first time that a group of major fishing companies has effectively created its own no-fishing zone. Perhaps significantly, the signatories include the giant state-owned China National Fisheries Corporation. China has by far the world's largest fishing fleet, estimated at 17,000 vessels, and in the past has had a bad reputation for plundering the oceans.

Promises are not the same as legislation, but this does suggest a change in approach among some fishing companies. Around a third of the world's fish stocks remain overfished. But nearly half of the fish caught worldwide are coming from fish populations that are increasing through more effective management, albeit often from a very low level. Optimism is growing. The idea that fish stocks are doomed 'is totally wrong', concluded Ray Hilborn of the University of Washington, in his academic paper analysing more than 800 separate stocks. 'Fish stocks are increasing in many places.'

But not everywhere. Rampant overfishing remains rife in the waters off much of Asia, for instance. Often, what is left are fish fit only to render down into fishmeal or oil for aquaculture. Hilborn notes, too, that some stocks destroyed during the bad old days are not recovering. They include Newfoundland cod, the pride of the North Atlantic until their numbers nosedived in 1992, herring in the northeast Atlantic, sardines off the US west coast, and Peruvian anchovies. Still, the omens are improving. The total weight of fish in the oceans was falling until 1995, but has been rising since 2005, he says.

Fisheries scientists such as Hilborn tend to be more optimistic about the state of the oceans than marine biologists, who are concerned about whole ecosystems, not just counting fish. Some have called Hilborn an 'overfishing denier'. But his detailed analysis is encouraging. And there is growing confidence, even among respected marine biologists, that the oceans can be restored to their former health by 2050.

'We have a narrow window of opportunity to deliver a healthy ocean for our grandchildren, and we have the knowledge and tools to do so,' Carlos Duarte, a marine ecologist at the King Abdullah University of Science and Technology in Saudi Arabia, was quoted as saying in the *Guardian*. He is lead author of a detailed review of the health of marine ecosystems that involved scientists from sixteen universities on four continents. It found that fishing was moderating, and the outright destruction of the coastal ecosystems that nurture most of the oceans' life was drawing to a close. It is not just fish species that are recovering. From Canadian sea otters to grey seals in the Baltic, from green turtles off Japan to elephant seals in the United States and great whales cruising the oceans, the news is generally good. 'Substantial recovery of the abundance, structure and function of marine life could be achieved by 2050, if major pressures – including climate change – are mitigated.'

Others agree. 'We can turn the oceans around. If you stop killing sea life and start protecting it, then it does come back,' Callum Roberts, now at the University of Exeter, told the *Guardian*. Historical data shows recovery often happens fast. During the First and Second World Wars, when marine warfare made fishing all but impossible, fish stocks recovered dramatically. In an interview with *Canadian Geographic*, Boris Worm, of Dalhousie University in Canada, admits to growing optimism. Recent history, not least the loss of cod stocks near his campus, gave many 'a sinking feeling that the ocean was permanently damaged. However in recent years, many of us have observed more and more signs of hope. We see the ocean display a remarkable resilience.' Few ocean species have gone extinct, so 'we still have all the pieces to rebuild the whole'.

Absolutely critical to the recovery of the oceans is the

fate of coastal ecosystems that nurture most marine life. Coral reefs and mangroves, kelp forests and seagrasses were lost on a huge scale in the late twentieth century. Mangrove swamps were replaced with prawn ponds, coral reefs were dynamited by fishers, dock builders ripped up seagrasses and much else. But here, too, there is good news, says Worm. He has seen how coral reefs in Indonesia have recovered after being repeatedly dynamited by fishers. 'You can still see where the bombs were thrown and you still see the rubble, but from that rubble emerges a new reef at a speed and an abundance that I frankly did not expect,' he says.

Other researchers, such as Virginie Duvat of the French research agency CNRS and Gerd Masselink of the University of Plymouth, are finding that healthy coral reefs can grow upwards fast enough to keep pace with rising sea levels. The Maldives, which is largely made up of hundreds of coral islands less than a metre high, has often been predicted to disappear beneath the waves by later this century. But so far, according to detailed analysis of NASA satellite images taken over the past four decades, many reefs are keeping pace with the current rates of rising tides, and the islands that they surround show no overall loss of land area. The same is true in the British-run Chagos Archipelago of the Indian Ocean.

This won't always happen, especially if sea-level rise accelerates, but 'the key thing to understand is that these islands aren't static', says Murray Ford of the University of Auckland, in the press release for his paper on the topic. 'They don't sit passively as if they were in a bath tub, slowly drowning.' But to fight back against rising tides, coral has to be healthy. Ford says building sea walls might seem an obvious way to protect low-lying islands from rising sea levels. But it could be the worse response, because the concrete will cut the coral off from the oceans that sustain its natural vigour and ability to grow. The evidence bears him out. Most of the coral islands lost in recent years to rising sea levels have first been damaged by human activity. Among the most famous losses have been coral atolls in the Carteret Islands off Papua New Guinea, most of which first had their coral reefs dynamited by fishers.

Even before rising tides, the bleaching of coral caused by rising ocean temperatures is a major threat to the ocean's reefs. It could destroy them all within decades. But a study of how 200 reefs responded after bleaching episodes found that healthy coral recovered more quickly, while pollution and overfishing make it more vulnerable to warming. It is a mistake to assume that coral reefs are doomed because of climate change, study author Mary Donovan of Arizona State University says in the press release. 'Local action to conserve coral reefs can help reefs withstand the effects of climate change.'

We can help in other ways. People are planting coral in damaged reefs from Costa Rica and Puerto Rico to Australia's Great Barrier Reef and the Cocos Islands. The technique is akin to grafting by gardeners. It involves snapping off small pieces of coral from existing reefs, and growing them on in protected marine nurseries, before transplanting them back to the reef. On Australia's Great Barrier Reef, coral gardeners are selecting species resistant to bleaching and other stresses.

In extremis, you can create artificial reefs. Shipwrecks and other marine junk create inert skeleton reefs on which live coral will eventually grow. Marine biologist Thomas Goreau and engineer Wolf Hilbertz came up with the idea of speeding the process up by running an electric current through seawater around these artificial reefs. The electricity mimics the processes the coral themselves use to grow, by causing minerals dissolved in the water to accumulate on the reef. The result is a limestone base to which live coral will migrate and grow.

The pair created their first electrified reef in the 1990s in the Negril Marine Park off Goreau's native Jamaica. 'We found that we were able to grow corals at three to five times the record rates, in a habitat where all the corals had been killed by pollution,' according to Goreau, in an interview with BBC News. Today they have more than 700 electric reefs growing coral, nurturing fisheries and protecting coastlines, most of them in Indonesia and the Marshall Islands.

• • •

Other coastal ecosystems can be restored too. Bob Orth of the Virginia Institute of Marine Science masterminds the world's largest seagrass restoration project in a bay on the Atlantic shores of the American state of Virginia. He often dons his snorkel and dives in to check on the progress of the 2,000 hectares of eelgrass – a type of seagrass – that have been planted in the past half-century since he started work there. 'They are like the prairies of the sea', he says.

Once, seagrass meadows were so profuse in the bay that

locals waded in to cut it for insulation material in local homes and schools. But in the early twentieth century storms, disease, sediment and pollution gradually reduced them. By the 1990s the bay was 'degraded and devoid of life', Orth says. Hope of recovery seemed distant, until locals spotted seeds of a local eelgrass floating in the water offshore. But the seeds were not reaching the bay. So Orth and his colleagues began collecting eelgrass seeds from nearby Chesapeake Bay and planting them. So far they have planted some 75 million seeds, Orth says. Today there are rays, seahorses and scallops swimming in the lustrous meadows, and migrating birds visit to join the offshore feast.

Seagrasses are found widely around the world's shorelines. Some marine biologists think we might protect and nurture them better if they had even more value to us, not just for sustaining fisheries but as a resource in their own right. Enter Spanish chef Ángel León. He has won Michelin stars for serving little-known and underappreciated seafood in his restaurant in Cadiz on the Atlantic shores of southwest Spain. He swam in the sea off the town for many years before noticing that the eelgrass that flourished there carried tiny green grains. He wondered if they might be edible. Research uncovered that eelgrass grains were a traditional food of the Seri indigenous people of the Gulf of California in Mexico. So he got cooking.

He now cultivates the grasses in a small marine garden, and serves the organic, protein-rich and gluten-free grains on his menu as the 'rice of the sea'. Leon believes his culinary innovation, which has been applauded by Orth, is of global importance. He has found a new source of food that, if gardened sensibly so the grass regrows, could provide an economic incentive for restoring one of the world's most important and endangered coastal ecosystems.

• • •

Like seagrasses, mangroves have been lost almost by accident, where communities across the tropics have chopped them down for coastal development and to make room for saline ponds to farm fish and prawns. But the impacts of these clearances have extended far beyond the loss of timber and of the crabs and fish that once lived among mangrove roots. Without the protective barrier of the mangroves and the mud in which they grew, entire coastlines have been lost to invading ocean waters. The prawn ponds have been washed away, and sometimes rice fields and whole villages too.

But now the tide may be turning, as coastal communities recognise the error of their ways and begin to restore the lost mangroves. On the coast of northern Sumatra, an Indonesian island left unprotected during the deadly Indian Ocean tsunami in 2004, more than 2 million new mangroves have been planted. In the village of Krueng Tunong – where more than a thousand bodies were found after the tsunami – Wahab has headed a community planting programme that has restored mangrove on 20 hectares around village ponds. 'We get more fish now that there are mangroves again,' he says. 'They grow faster and in greater numbers than when the ponds were bare. The roots help them avoid predators. We get more crabs, too.'

On the other side of the world, on the Pacific Ocean

shore of El Salvador, Carmen Argueta takes village women on mangrove planting days in the Bay of Jiquilisco. In an afternoon, they may plant in the mud 300 or more seedlings they raise in a village nursery. This initiative, too, began after a natural disaster, when Hurricane Mitch blew through in 1998. 'Afterwards, we realised that the areas with healthier, bigger mangroves were not damaged as much,' says Carmen. 'That got us to thinking.'

Mangroves in their bay can live to sixty years old and grow to 70 metres high. 'They act as natural barriers against tsunamis, floods or other natural phenomena,' she says. Healthy mangrove roots, which are partly in the mud, partly in the water and partly exposed to the air, 'help protect against rising water'. No other trees are like this. 'They also provide healthier water for small fish to grow. Communities can fish more, which improves the economy for families. Mangroves are life for our generation and for future generations. If we neglect them, we neglect ourselves.'

After the hurricane, Carmen and her friends set up a local association to make their village safe by protecting the bay's remaining mangroves and planting more. So far, they have restored some 200 hectares. 'There is a great communal feeling when we do this,' she says. 'We get in the water up to our waists, putting sediment into buckets. People don't mind, because they are doing an important job for their families and for their ecosystems.'

To start with, the success rate for planting was low. 'We didn't know enough about the right places to plant. But we found experts in Thailand, who taught us.' Now, the association's rangers keep track of when their mangroves produce seedlings that can be picked and replanted to extend the mangrove areas. They are methodical, picking the right species to plant at the right density in the right area, Carmen says. Another trick is to regularly remove garbage and dead branches from the water channels that connect the sea and river waters. This maintains the mix of saline and fresh water vital to the mangroves.

And increasingly they avoid planting altogether in favour of helping seedlings to regenerate naturally. This they do by putting wooden fences in the mud to catch silt and elevate the land. The results are impressive. Fish now dart through the mangrove roots while birds sing in the branches, and animal tracks are visible in the mud. Under the rules of the biosphere reserve where they live, only villagers are allowed to enter the mangroves, fishing, hunting crabs, cutting wood and collecting shellfish. El Salvador has lost 60 per cent of its mangroves in the past half-century, but Carmen says this is the fight back.

• • •

The NGO Wetlands International reckons worldwide most mangrove plantings fail because not enough attention is given to what to plant and where. A study of planting programmes in Sri Lanka, which has the richest mangrove biodiversity anywhere, found that on twenty out of twenty-three sites looked at, most trees died. At nine sites, inspectors could find no surviving plants at all. A better idea is to create the conditions where nature will reseed mangroves. Given the chance, mangroves will grow almost anywhere in the tropics. But they require a muddy shoreline, and tides and currents that are slack enough to allow seedlings to take root and grow.

They will often return to abandoned prawn ponds or rice paddy. Guinea-Bissau in West Africa is the most mangrove-covered country in the world. For some decades, the country lost mangroves as low-lying coastal areas were taken over for growing rice. But recently, young people have been moving to the towns, abandoning their fields and letting the dykes around them fall into disrepair. Result: return of the mangroves. Now environmentalists are working with villagers to remove the remaining dykes and restore yet more.

On the north shore of the Indonesian island of Java, villagers a generation ago followed the fad of removing mangroves to make way for prawn ponds. But the sea invaded, engulfing the ponds and rice fields inland, drowning villages and leaving others held up on stilts and surrounded by water. But they are fighting back, by restoring the mangroves.

To reach Timbulsloko, you have to drive 7 kilometres along a narrow causeway passing half-submerged houses and a village cemetery where lapping waters occasionally float decomposing bodies into nearby living rooms. At the far end of the causeway, in the village hall, elders remember the days before the flood. 'I grew up in the 1960s when the sea was more than a kilometre away,' says Slamet, interviewed for a report for Wetlands International by one of the authors. 'Then, after we cut the mangroves, the waves began to get bigger.'

Village leader Mat Sairi was honest in the report about

the mistakes they made back then. 'Our parents told us we should protect the mangroves,' he remembers. 'They said they provided many benefits, like the oysters, crabs and fish among their roots as well as protection of the coastline. But our people wanted the land to make money and feed their families.'

Around 19 square kilometres of land along this stretch of coastline have been flooded in recent years because of mangrove loss. Many residents of Timbulsloko and its neighbouring villages left. But Mat and his fellow villagers are intent on recovering their land. Beyond the end of the causeway, in an area of shallow water that was once land, they are erecting long brushwood barriers where the old coastline was. The structures are 175 metres long and rise more than a metre above the waves.

The purpose of these strange constructions – like giant nets on a tennis court – is to still the waves enough to prevent further erosion and trap silt where mangrove seedlings from elsewhere along the coast can settle and grow. 'Mangroves require a stable coast to settle,' says Apri Susanto Astra of Wetlands International in the report. Without still waters, planted mangroves usually get washed away.

All told, the structures extend along 14 kilometres of the coastline. Mat tours his structures. Reaching the first barrier, he uses a paddle to check the depth of the water. The barrier, in place for only eight months, has raised the level of the mud on the landward side of the barrier by 15 centimetres. Soon after construction, the first mangrove seedlings poked their shoots above the water. 'We are not leaving,' he says. 'This is our home and, God willing, we plan to stay.'

• • •

All across the tropics, they are realising how disastrous it was to clear away mangroves for prawn farms. Those ponds that weren't invaded by the ocean have usually seen prawn output crash as diseases take hold. And governments anxious to protect their coastlines from disappearing into the sea are banning the clearance of more mangroves.

Nonetheless, other forms of aquaculture continue to take over coastlines around the world – from salmon cages in Scandinavian fjords to catfish ponds across the Mekong delta, and oyster beds in Florida to carp pens off the coast of China. These open-water fish farms may not directly displace mangroves, but they can damage seagrasses and coral reefs, and pollute the ocean with fish excreta, undigested fish feed and antibiotics. The fish can also damage wild fish populations if they escape from their pens and compete for food or interbreed with their hosts.

Still, marine farming has a role to play. The world has to be fed, and aquaculture is the world's fastest-growing source of food. With wild fish catches either declining or curtailed to protect stocks, aquaculture production may before long exceed wild fish catches. But the two are not at present separate. Like livestock on land farms, the fish in fish farms need feeding – and mostly they are being fed on wild fish. China alone catches up to 4 million tonnes of wild fish to feed to fish in its aquaculture pens.

Some environment groups say now is the moment to begin embracing aquaculture and to begin a long march to make it sustainable. The Nature Conservancy (TNC), an influential US-based group, says aquaculture has a lot going for it. Most farmed fish have a much smaller environmental footprint than livestock on land. Salmon can convert almost all their feed into new protein, whereas the ratio for beef cattle is eight to one. 'Done well, [aquaculture] can be a force for ecological and social good,' says Robert Jones, global lead for TNC's aquaculture programme, in a TNC report. He wants a 'blue revolution' to catalyse private investment in sustainable aquaculture. 'That is easier said than done,' says Richard Waite in a report from the World Resources Institute, a US-based think tank. 'To date, intensification has led to an increase in the use of energy and fish-based feed ingredients, as well as an increase in water pollution.'

To do aquaculture better, the first need is to move fish farming offshore, away from fragile coastal environments, to places where no coastal ecosystems are at risk and where pollution can disperse on the currents. Big cages at sea can also maximise productivity, delivering up to 500 tonnes of fish per hectare each year. Optimists say that offshore fish farming could even sometimes bring gains for marine life. 'A wide variety of fauna live or gather in and around mariculture farms,' says Rebecca Gentry of Florida State University in a paper on marine aquaculture.

A second offshore opportunity is culturing shellfish such as oysters, clams and mussels, as well as seaweed, which is finding new markets in cosmetics, animal feed and biofuels.

Advocates say these crops can also bring benefits to ecosystems. Shellfish beds provide habitat for other creatures and mop up pollution, while seaweed production can neutralise dead zones by absorbing nitrogen pollution from rivers, and can fight climate change by taking up carbon dioxide.

The seabed itself could be re-engineered for offshore aquaculture. One approach sinks artificial reefs to encourage the growth of marine life, on which fish can feed. Such reefs could be open to the ocean, allowing marine life to come and go, or giant enclosed cages stocked with farmed juvenile fish. The Chinese Ministry of Agriculture has announced plans for more than a hundred such new marine ranches, covering 2,700 square kilometres of ocean floor and involving 50 million cubic metres of artificial reefs. But there are concerns. In an interview with one of the authors for *China Dialogue*, Songlin Wang of the Aquaculture Stewardship Council, a certifying body, says one such project 'is planned on top of a seagrass bed that is a valuable fish habitat and biodiversity hot spot'.

Still, Chinese-style offshore aquaculture could become the marine equivalent of European vertical farms, perhaps leaving the rest of the ocean to recover, says Neil Sims of aquaculture pioneers Ocean Era. His farms operate in the open ocean off his home there, and also a kilometre off Mexico, where he says strong currents and deeper waters dilute waste, protecting coastlines. Neil has also tried unmoored pens that drift with the currents, which he hopes will reduce environmental impacts further. One is attached to a barge carrying feed for automatic delivery to its 2,000 kampachi fish. Technicians run the farm remotely by phone, visiting the giant cage just once a week to top up the feeder.

Sims sees such systems as a model for a massive scale-up of offshore aquaculture. 'We could produce a hundred times the current level of seafood,' he says. 'It would be fish farming that actually supports the recovery of our oceans, rather than hinders it.'

• • •

The payback for successfully restoring the oceans would extend far beyond fisheries or even marine ecology. There would be a big carbon dividend. The restoration of coastal ecosystems is a great way of capturing carbon dioxide from the air to slow down climate change. These ecosystems typically take up carbon dioxide several times faster than rainforests – with mangroves and seagrasses doing the best. Some call it 'blue carbon'. By some estimates, just stopping the loss of coastal ecosystems would keep half a billion tonnes of carbon dioxide out of the atmosphere each year. Putting them back would achieve the holy grail for climate scientists of 'negative emissions', of which we will hear more in a later chapter.

Restoring the oceans would also speed up the return of the planet's great whales. That would have an additional huge benefit for the ocean's ability to soak up carbon dioxide, promises marine biologist Joe Roman of the University of Vermont. His research shows that whales are great ocean gardeners. They stir up the water, bringing nutrients and minerals to the surface. Their poop fertilises surface waters, which in turn can sustain more plankton as food for marine life, including the whales themselves. More plankton further increases the marine ecosystem's uptake of carbon dioxide from the air. Wherever whales go, they increase plankton production, so wherever they go the oceans can take up more carbon dioxide too. Roman calculates that one grey whale captures around 30 tonnes of carbon in its life, the equivalent of more than a hundred trees. When the whale dies, it takes that carbon to the seabed.

Hiram Rosales-Nanduca of the Autonomous University of Baja California, in Mexico, who has studied the whales swimming near his lab for more than two decades, sees them as the ocean's great movers and shakers.

> They are the biggest animals in the history of our planet, some of them, like the bowhead whale, live to be more than 200 years old, and they are engineers of the oceans. If we have more whales, thanks to conservation, we have more plankton and more capacity to remove carbon from the atmosphere. This is essential and maybe the whales can make the difference.

The world talks a lot about finding ways to capture carbon dioxide from the air and bury it out of harm's way. Increasingly, the focus is on engineering solutions. Well, given the chance, whales will do the engineering for free. Economist Ralph Chami of the International Monetary Fund (IMF), in an article for the IMF website, has tried calculating the value of whales as machines for capturing carbon. He reckons the average great whale is bankable to the tune of more than $2 million. That puts the current total carbon value of the world's great whales at more than a trillion dollars. Saving whales should be at the heart of the objectives of governments signed up to the Paris climate agreement, says Chami.

Whales are in recovery today. Most boats chasing them are full of tourists rather than men with harpoons. Since the international moratorium on hunting was agreed in 1986, their numbers have risen into the hundreds of thousands. Some species, such as North Pacific humpbacks and southern right whales, are nearly back to their levels before industrial hunting. Others, such as blue whales, are taking their time, while still others may never come back, because the right conditions are disappearing. For instance, global warming means that the area of cold waters that Arctic bowhead whales need will have halved by later this century, says Roman.

Still, recovery is happening. Whales live a long time and breed only slowly, so this will inevitably take time. But it is remarkable that they are probably now back to about a fifth of their former numbers. The return of the great whales is a symbol both of the improving health of the oceans and of how much more there is to be done. 'Healthy whale populations imply healthy marine life including fish, seabirds and the overall vibrant system that recycles nutrients between oceans and land, improving life in both places,' says Chami. He calls it 'earth tech'. But it is also earth life.

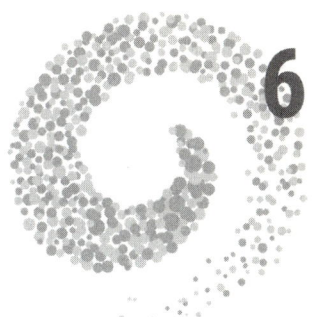

Clean Our Air

S pace is a great teacher. It teaches you equality, resilience and patience, and it teaches you about how to live on your own planet.

I was born a year after the Apollo 11 Moon landing, and I think it was fate that I grew up wanting to be an astronaut. When I was little I lived near Tokyo and visited the planetarium with my parents and older brother so I could learn more about the stars and the constellations. But it was always very difficult to see the stars at night in the city. Then, when I was five years old, we moved to Hokkaido, the northern island. There was no pollution there, and the sky was so clear and beautiful. I was able to see stars in the night sky for the first time. At school we learned that human beings are made of almost exactly the same things as stars and planets: oxygen, hydrogen and nitrogen. We humans are siblings of these stars and the planets, and part of the same universe.

Before I became an astronaut, I always thought

space was the most incredible place, and the most desirable place to get to. But when I first saw our home planet through the windows of the space shuttle I suddenly thought, oh, maybe it's the Earth that is that special place; unique in the vastness of the universe. I could see the shining blue of the ocean, the clouds moving during the day. I felt like it was alive, and it was more beautiful than I had imagined.

From space I could also literally see the air we breathe – a delicate thin blue line of atmosphere – and in that moment I realised how precious our atmosphere is. Just like the supply that sustains us in the shuttle, that thin line is our only oxygen supply. Whether you're in space or on Earth, air is a finite resource. Against the vastness of space it's just this paper-thin atmosphere that separates us from an airless world. So we are very lucky.

But from a distance you can also clearly see the damage that humanity has brought on our home. You can see forest fires, deforestation, impure water and air pollution over industrial areas, natural disasters. The ground feels so steady under our feet, but when you look from the outside you can see how fragile our planet really is. But if we have the determination and skills to carry people like me into space, then we can certainly clean our air too.

A thing about astronauts is, we are all rivals; everyone wants to go to space first. But another thing we know is that we must support each other, not only to further space exploration, but for a purely practical reason – to survive. In the space station, if the air was contaminated or full of chemicals, it would be a disaster, and we would have to do everything in our power to clean it because our lives depend on it. We know today that the air on our planet is polluted, so we need to act with the same urgency to clean it – our lives depend on it. Japanese culture combines a deep respect for the natural world with a love of technology and innovation. Harnessing the power of nature, technology and innovation to help clean our air is what we must do now, but on a global scale.

Our planet is our spacecraft, and we are all members of its crew, each carrying our own unique role and responsibility. If we unite our strengths and share our information, we can defeat our global challenges. Just imagine the future we could have if we choose to do so – a world with clean, safe air, where every child can see stars.

Naoko Yamazaki, 2021

From the smog clouds in our cities to the dark snow in Scandinavia, from the thinning ozone layer to plastic particles falling out over the oceans, air pollution is almost everywhere. Nobody is safe. The majority of the world's population breathes dangerously dirty air every day. Nobody thinks this is acceptable. In 2019 the UN launched the Clean Air Initiative, calling on governments across the world to achieve air quality that is safe for their citizens within the next decade. But doing that will require treating our shared atmosphere as a priceless, finite resource to be protected for all.

So how can we purify our city air? How can we cleanse the countryside and keep our oceans fallout free? How can we restore the chemical balance that the atmosphere on Earth has maintained for thousands of years?

The great clean-up will sometimes require technical breakthroughs or the adoption of new international treaties. But often much of the death toll is down to lax standards, poor policing of existing laws, failure to adopt cheap and easily available clean-up technology and sheer indifference – failures that could be fixed almost overnight: ending the burning of coal in cities; builders spraying water to suppress dust on construction sites; ships putting filters on their funnels; the dirtiest planes, trucks and cars retired as soon as possible.

This chapter will chart the good and the bad, the unlikely victories and appalling failures, what is too late to prevent and what we urgently need to halt today, where technology is needed and where ignorance and lassitude must be expunged so the existing technology can do its job. But whether the need is for innovation, adoption, regulation or simple common sense, there is much to do to address airborne problems that often seem to slip below our radar.

• • •

There are some success stories to tell. We banished coal burning in many cities back in the twentieth century, ending London's pea-souper smogs and much else. For hundreds of years London had been known as the 'big smoke' because fumes from coal fires filled its streets for weeks on end. Those days are gone. Then acid rain was largely ended in Europe and North America by adding sulphur-capturing equipment to power station smoke stacks – ending a holocaust that had wiped fish from thousands of lakes, acidified soils and sickened millions of trees. Perhaps most impressive of all, the chemicals that rise up through the atmosphere and eat the ozone layer have been largely banished, though the ozone layer itself will take decades to fully recover.

But those were twentieth-century solutions to twentieth-century problems. Now in the twenty-first century we are increasingly aware of modern invisible urban air pollution

Polluted skies over Shanghai. In cities around the world we now know our air quality is leading to major health issues for humans as well as nature; but we've also seen how huge strides can be made to clean up our skies.

After the Chernobyl nuclear disaster scientists were amazed by how quickly plants and animals recolonised the city.

Wildebeest crossing the Mara River. Many of the great migrations are being impacted by climate change and the breaking up of animals' routes by agriculture or deforestation. This migration is now one of just a handful of great migrations still occurring.

A controlled burn for land clearance in Madagascar. Across the world we are losing our forests primarily to provide land for agriculture.

The Noor Power Plant in Morocco is the size of San Francisco and demonstrates the vast potential for solar power across the desert regions of the world.

The sea freezing in Ittoqqortoormiit, East Greenland; something that is happening later and later each year.

Polar bears are the Arctic's top predator. They rely on sea ice for hunting, resting, travelling and even sometimes for dens. As the sea ice declines all of their activities are impacted and as they spend more time on land polar bears come into more frequent conflict with people.

Reindeer herding is one of the most ancient examples of how animals and humans can work together, but their migratory routes are increasingly affected by climate change and their food sources by air pollution.

A polluted riverside in Ho Chi Minh city during a flood tide. Our waste is often transported around the world – sometimes unintentionally through ocean currents or rivers and sometimes as part of waste management projects. Both can lead to huge volumes of waste from Western countries ending up on the other side of the world.

Singapore is pioneering a form of green building that incorporates nature throughout. The result is having a beneficial impact on air quality, temperature and the wellbeing of the people that live or work in those buildings.

A soybean plantation (left) meeting the start of a protected area of Brazilian forest. Soybean agriculture is one of the biggest causes of deforestation worldwide. It is often fed to chickens and cows.

A high tech farm in Holland that aims to minimise the inputs it uses and maximise yield.

A male orangutan in a rehabilitation centre in Borneo, where much of their habitat has been destroyed for palm oil plantations. Increasingly pioneering farmers are trying to find ways to grow palm oil only on land cleared long ago and recreate corridors to connect the remaining intact forest.

from tiny particles and nitrogen oxide gases that are causing epidemics of asthma and other diseases of hearts and lungs. They kill millions each year.

But we also have twenty-first-century solutions. Step forward what will probably turn out to be the most dramatic of them all: the electric car. The end of the internal combustion engine may be in sight. Electric cars are going to transform our world. We will wonder, as hospitals empty of patients with crippled lungs and over-worked hearts, why we didn't find a way to go electric decades ago.

The rise of the electric vehicle has been dramatic. A decade ago, they barely existed. By 2021 there were 11 million of them on the roads, plus approaching a million buses and trucks. That is less than 1 per cent of all vehicles, but in 2021 5 per cent of new cars being bought worldwide were electric. The race is on, and it is a race to the top, to a clean-air future. Big car manufacturers are no longer trying to hold back the new; they are rushing to dominate the market. Virtually all the world's large vehicle manufac-turers are ramping up new models, establishing new production lines and promising they will have gone entirely electric in a decade or so. The rise of electric vehicles is proving even more rapid that the rise over the past decade of solar and wind energy.

The International Energy Agency (IEA), once a conser-vative defender of all kinds of fossil fuel-burning technologies, says electric vehicles can and must take over our roads. In a road map to get the world to zero emissions of carbon dioxide, published in early 2021, it anticipates that 60 per cent of all new cars will be electric worldwide by 2030.

The prime policy incentive for this transformation is the cutting of carbon dioxide emissions. How good electric vehicles will be for the climate depends almost entirely on how the electricity they require is generated. If it is made from burning coal, oil or gas, then the climate benefits are minimal. If it is from wind and solar power, then the gains will be huge in future decades, as we will discuss in the next chapter.

But their first and most obvious impact will not be on global climate but on the air in our streets. No internal combustion engine means no burning of fuel. That means no pollution coming out of exhaust pipes. No nitrogen oxides, no carbon monoxide, no hydrocarbons, no tiny lung-penetrating diesel particles. No ozone smogs cooked up in summer heat. No winter haze. Instantly, car by car, our street air will clean. And our children's lungs will breathe easy.

Until recently, electric cars were seen as city run-arounds for the rich, with only a limited range before they needed plugging in. Now that range is increasing fast. Beyond 400 kilometres before a recharge is necessary. And super-fast charging points are springing up everywhere. In 2021 the British government announced a £300 million investment in points that can fully recharge a car battery in thirty minutes.

Amol Phadke of the University of California, Berkeley, says all new cars sold in the United States will be electric by 2035. General Motors say they will meet that date for sure. We may go even faster. We should. But almost certainly, by 2040 the only non-electric cars on the roads will be an ageing fleet of vintage rust buckets on specially licensed rallies – the steam engines of the roads. Hybrids too will be long gone, replaced by fully battery-powered vehicles.

We will be able to smell the difference. With nitrogen oxide emissions from all sources down by 85 per cent and the finest particles reduced by 90 per cent, the air will be sweeter, and hugely healthier. Cities will be places people want to live in by choice, not necessity. The switch to electric vehicles will prevent at least 2 million premature deaths from air pollution, reckons the IEA.

There is no doubt that a revolution is under way, and some people are getting very rich as a result. California entrepreneur Elon Musk, whose company Tesla pioneered modern electric cars, had by early 2021 become the richest man on the planet, personally valued at almost $200 billion. His story shows that green technology has become a smart way to wealth. Why be a banker when you can be a green technologist?

• • •

Most of us presume that air pollution comes largely from cities, and spreads out to the countryside on the winds. But one of the most insidious pollutants starts in the countryside, on farms, and often spreads to cities, where it adds to urban smogs. The perverse pollutant is ammonia.

Probably, you didn't know ammonia was a problem;

certainly not down on the farm. The problem is slurry tanks in big poultry sheds, pig pens or cattle feeding lots, where urine and faeces mix. The chemical reactions between the two wastes generate large amounts of the gas, which bubbles off into the air. The precise output of ammonia depends on a lot of environmental conditions in the animal sheds and slurry tanks. But the bottom line is that what was once a source of valuable fertiliser for pastures is, on factory farms, a waste with a toxic by-product. The ammonia floats away on the winds, and can fall out far away, killing ecosystems – or mixing with urban pollution and killing people.

Britain is one of the most seriously impacted countries. A single dairy cow produces 40 litres of urine a day. With more than 9 million cows in the country, that is a lot of urine and potentially a lot of ammonia. The Scottish Environment Protection Agency found some individual poultry farms emit hundreds of tonnes of ammonia into the air annually. Overall, more than a quarter of a million tonnes of ammonia escape into the air from British farms every year. Research aircraft have recorded a tenfold increase in airborne concentrations of the gas as they pass over farms.

Ammonia fallout back to the ground, often in 'ammonia rain', can exceed 20 kilograms per hectare per year. It is like a giant extra dose of fertiliser. This is bad for the many ecosystems that flourish in clean air and nutrient-poor soils. Added doses of ammonia have been found to reduce the biodiversity in species hot spots ranging from southern China to eastern Brazil. In the Netherlands, it is changing heaths into grasslands. In Britain, where more than 85 per cent of the country receives ammonia above critical levels for sensitive species, the fallout is killing lichens, mosses and fungi, and damaging the majority of sites of special scientific interest, turning heather moorlands to grass.

The threat is not just to wildlife, or to the countryside. Ammonia drifting from agricultural areas to towns reacts with other pollutants in urban air to create fine particles that we breathe deep into our lungs. Agriculture is the single biggest contributor to particulate levels in some British cities, including Edinburgh, says ecologist Mark Sutton, who works there. Researchers reckon 3,000 British lives could be saved each year by halving farm ammonia emissions. Globally, the figure may be 250,000.

Halving emissions would not be hard. Some countries are showing the way. In the Netherlands, where cattle and pigs outnumber people, they have cut emissions by two-thirds since 1990. But many governments, including the UK, do not routinely monitor ammonia emissions from individual farms, and have no clear policy for cutting its deadly fallout. British ammonia emissions have barely changed in recent years, says Sutton, and regulations that should require farmers to clean up go largely unenforced.

There are plenty of ways to cut emissions. No new tech is required, just extra care. Keeping slurry tanks cool would help. So would covering them to prevent the release of ammonia. And where slurry is put on to fields, it should be injected into soils rather than sprayed into the air. Even planting shelter belts of trees and hedges nearby would help to absorb escaping fumes.

But the real answer, say many, lies in curtailing the hundreds of giant factory farms housing millions of pigs, poultry and cattle that accumulate large volumes of ammonia-producing slurry with nowhere to go. When most livestock still grazed fields, their urine and faeces returned to the land to fertilise more grass for the animals to eat. Today it is a waste stream that causes pollution and death. It may not be possible, or sensible, to go back to universal grazing. But what comes out of their back ends needs to become a resource once again.

• • •

Making our air safe again will take many forms. One little-discussed topic will require some ingenious technology. Perhaps that technology has already been cooked up by some students brainstorming after lectures at Imperial College London.

It is becoming clear that while electric cars will transform the smell and safety of our city streets, they won't end all air pollution from vehicles. Up to half the particles that penetrate deep into our lungs as we breathe city air do not come from the exhaust pipes of vehicles. They come from another pollution source you have probably never considered before: vehicle tyres. Specifically, from the tiny bits of rubber, plastic and chemicals that fly off whenever vehicles brake, accelerate or turn. When the rubber hits the road, there is a direct pollution pay-off.

Engineering student Hugo Richardson hadn't heard about this hidden cause of air pollution until a teacher talking about microplastics at Imperial mentioned it in passing during a lecture. It was almost a throwaway remark, Hugo remembers. He didn't have much to say because very little actual research had been done. Afterwards, Hugo and some course-mates got talking. 'We thought: why have we never heard of this?' The four friends figured the air of their home cities, in the UK, the United States, India and China, was being saturated with tyre fragments. So they formed a group, called The Tyre Collective, to try to figure out what to do.

They found an obscure academic paper that had looked at rubber coming off the tyres of a bus on London's longest bus route, the X26 travelling 35 kilometres from Croydon to Heathrow. The bus made the journey eleven times a day, and in that time discharged into the air 300 grams of tiny tyre particles, enough to fill a container the size of a grapefruit. Not a lot, perhaps. But that was just one bus among 9,000 London buses and 2.5 million vehicles of all kinds in the city. 'Our first reaction was complete shock, and being quite disgusted in many ways,' Hugo says.

The more they looked, the scarier the issue seemed. Some UK government scientists had taken a brief look at the issue. According to a preliminary review by the government's air quality expert group, tyres contribute well over half of all emissions of small particles from vehicles – right up there with diesel particles. 'Yet no legislation is in place specifically to limit or reduce these particles,' said the expert group. More widely, half a million tonnes of particles fly off vehicle tyres across Europe each year. As diesel exhaust emissions are cleaned up, an ever growing proportion of what is left in the air comes from tyres.

How dangerous are these particles? Rubber is just one component of tyres, says Collective member Deepak Mallya. 'There are hundreds of different chemicals that go into making tyres, from sulphur to zinc to carbon black, so these chemicals react with different chemicals in our environment and become more toxic.' And if they are like diesel particles, they are killing millions every year from lung and heart diseases. In any case, their small size makes them inherently dangerous. They will get deep into our lungs, irritate the lung lining, and may cross into our bloodstream. Like asbestos. The World Health Organization says there is no safe amount of particles that size.

Besides the threat they pose to our health, tiny tyre fragments make up a large proportion of the microplastics blowing about in the air. In one study in the American west, they made up more than 80 per cent of airborne microplastics. They also appear in the cocktail of microplastics swirling around in the oceans. We have all heard about plastic bags and six-pack rings and fishing gear – but tyre wear is bigger than any of them.

After gasping at the scale of the problem, Hugo and the Collective decided being scared didn't help much, unless you had a solution. They soon discovered that nobody – not car makers, not tyre makers, not the people who run the roads, not even public health professionals – had seriously thought about this. They took that as a challenge. First they looked for an alternative, more durable material for making tyres, one that would produce fewer fragments.

But nothing they came up with seemed to be both practical and effective. So they figured the next best thing would be a device for catching the particles as they came off the tyre. A vacuum cleaner stuck behind the wheel, maybe? A better idea, they decided, would be a device using static electricity to attract them into a container.

Any child knows that rubbing a balloon against your jumper gives it an electric charge. In the same way, tyre fragments pick up a positive charge from the friction between the tyre and road. So the Collective put on their engineering hats, and came up with a device that has a small negatively charged plate, like an electrode on a battery, to attract the particles as they fly off. It then directs them into a small tank behind the wheel or in the wheel arch. The prototype they developed can collect about 60 per cent of the fragments, so there may be more refinement to be done.

Still, they are on the road, trying out the prototype on buses. Buses are ideal, because they have high emissions through constantly stopping and starting to let passengers on and off, and turning at junctions along their routes. They also have regular maintenance schedules for emptying the tanks of the tyre waste. If they can demonstrate success, they hope for orders. Meanwhile, they are also working to interest vehicle fleet companies and car manufacturers.

They foresee a world in which every tyre on every vehicle on every road has a microplastic collection device fitted as standard. They would be just as routine as the catalytic converters adopted in recent decades to capture some toxic gases from vehicle exhausts, says the American scientist among the collective members, Siobhan Anderson.

187

The urgency of the case for doing this should increase with the spread of electric cars, they believe. First, because as exhaust pipe emissions disappear so the importance of tyre fragments in total particle pollution will become increasingly obvious. And second, says Hugo, 'because electric cars are heavier, what with the batteries they must carry. So they will probably produce even more tyre particles.'

Next up, the Collective wants to find a reuse for the tyre particles that their device will collect – potentially thousands of tonnes of it every year. The rubbery residue could perhaps be recycled to make new tyres, or even the soles of shoes. For now, they have made their business cards from ink made from the tyre particles they harvest. Got any other good ideas?

• • •

Another major affliction of the air in many countries is fumes from brick production. When heated to around a thousand degrees, during the process of turning clay into bricks, brick kilns can produce toxic fumes. The world makes 1.5 trillion bricks a year in perhaps 100,000 kilns. It is a low-cost industry with virtually no regulation in many countries. Workers operate sometimes in near slavery conditions, living on site; and the bricks are made in primitive kilns with little or no effort made to prevent the smoke spreading across the landscape, and little or no outside regulation.

The countries that make the most are those with fast-growing urban areas and limited access to hard rocks or limestone to make cement. Brick making is probably the fastest-growing and least regulated industry in Asia. Bangladesh, India, Pakistan, Brazil and Mexico have some of the most intense production, and many of the most polluting kilns.

There are two kinds of pollution. One is fluoride from the clay. Few Britons remember that as late as the 1980s, a vast expanse of brickworks in the English East Midlands, around Bedford and Peterborough, emitted fluoride. Cows that grazed in fields near the brickworks were sometimes born deformed. Some died. Humans were reckoned to be at risk too. For while in tiny quantities fluoride prevents dental disease, in larger quantities it causes bone deformities.

Thousands of people in India are known to suffer such deformities from drinking water naturally contaminated by fluoride. But many more could be suffering undetected from the intense fallout of fluoride from local brickworks. One study, conducted near brickworks around the city of Lucknow, the capital of Uttar Pradesh, found fluoride levels in spinach, the most popular locally grown green vegetable, that averaged fifty times legal limits.

Ron Fuge of Aberystwyth University, in a paper in 2019 in *Applied Geochemistry*, estimates that approaching 2 million tonnes of fluoride is poured into the air annually from brickworks worldwide, making brickworks by far the largest airborne source of one of the world's most pernicious poisons. The problem is also among the least researched.

The other pollution concern with regards to brickworks is the smoke and fumes created by the large amounts of fuel used in kilns to bake the clay. Coal is the most common fuel, followed by wood, charcoal, animal dung, tyres and even plastic waste. Most producers in problem countries make little effort to clean the emissions, which simply go straight up the stacks, blackening the skies all around.

The Climate and Clean Air Coalition, an intergovernmental body hosted by the United Nations Environment Programme, reckons that brick kilns account for a staggering one-fifth of all the world's soot emissions. The black carbon particles penetrate the lungs of workers and people all around. They damage crops. In Bangladesh, they are blamed for declines in mango crops in the north of the country and rice in the south. And some of the soot spreads on the wind, eventually falling to the ground on distant glaciers and ice caps, causing them to melt more quickly.

Fixing this is not rocket science. Studies have found that the amount of fluoride released from baked clay depends a lot on the temperature in the heart of the kiln. Proper control to keep temperatures below a thousand degrees would eliminate 80 per cent of emissions, says Fuge. Reducing the soot requires putting soot-catching chambers and filters in kiln stacks. Eventually, coal and other dirty fuels should be replaced with natural gas. Just making the kilns more efficient would reduce pollution by up to 50 per cent, says the Climate and Clean Air Coalition.

China has massively reduced its brick kiln emissions in the past decade by setting and enforcing standards for technology and emissions. Its brick production has gone from being one of the most polluting to the least polluting in the developing world. Other countries have no excuse not to follow its lead.

. . .

The good news is that when governments are persuaded of the need to act, they often do so speedily and effectively. Little more than thirty years ago, the rapidly growing hole in the ozone layer was seen as the number one environmental issue for the world. It loomed large as an imminent global disaster at a time when global warming was seen as a distant and only theoretical threat. Chemicals known as chlorofluorocarbons (CFCs) used in refrigeration and aerosols sprays were discovered in the mid 1980s to be causing the rapid destruction of the ozone layer, especially over Antarctica. Ultraviolet radiation from the sun was streaming through the thinned layer and was being blamed for widespread rises in skin cancers in Australia and elsewhere. Sea life might be threatened. It was a threat that nobody had suspected, and which spooked scientists.

In some panic, just two years after the discovery a deal was reached at a conference in Montreal in 1987 that banned further manufacture of CFCs, while subsequent deals have dealt with other chemicals with a similar if smaller effect. Manufacturers rapidly switched to alternatives. There have been problems since in ensuring that the CFCs in tens of millions of refrigerators around the world were not released into the air when the equipment reached the end of its working life. Warehouses filled with fridges awaiting safe emptying. Still, the Montreal Protocol did its job; CFCs emissions have largely ceased, little more is floating into the ozone layer over our heads, and as a result the ozone layer has begun to recover. Complete recovery will take decades yet. But the panic is over. At least one global life-threatening environmental problem has been fixed.

This episode shows what the world can do if it has a will. But it also shows that, for many sources of air pollution, it is vastly easier to staunch the flow at the source than to clean up afterwards. We are discovering this with many of the other 140,000 chemicals manufactured by humans during the twentieth century for use in consumer products, industrial processes, as part of electrical equipment or as pesticides. Many such chemicals are harmless, but some are poisonous. And once in widespread use, most end up escaping into the environment, whether evaporating from fields, discharged from factory chimneys or released from scrapped equipment.

Collectively, these chemicals are often known as persistent organic pollutants (POPs). Among the most persistent and most toxic are flame-retardant PCBs and pesticides such as DDT. Once they have escaped, POPs stick around, gradually accumulating in soils, water, plants and the bodies of animals. For a while nobody notices, but before long they reach concentrations where the damage they are doing becomes clear, as birds die, polar bears become emaciated or change gender, and freshwater ecosystems start losing predators at the top of the food chain, such as otters.

With growing public anger at the carnage caused by POPs, many industrial countries banned production in the final decades of the twentieth century. A formal international treaty was finally agreed in Stockholm in 2001, banning further production of some of the most troubling of them. But you cannot put these chemical genies back into the bottle. They live on in the atmosphere and oceans, constantly being recycled through evaporation into the air and rained out again. We cannot unrelease them. They are almost impossible to round up. We can only try to ensure that nothing similarly dangerous is approved for production and release in future.

• • •

While some problems from air pollution are all but impossible to fix, others can and should be dealt with. But somehow we fail in the task. If electric vehicles can transform the air in urban streets where half the world's population now live, we have an equally important task to clean up the air in rural villages and homes.

Across huge areas of the developing world – from the desert fringes of the Sahel to the migrant suburbs of Ulaanbataar in Mongolia, the million villages of India to the squatter camps of Mexico – cooking smoke is the most distinctive of all smells. As evening approaches and shadows lengthen, people in tens of millions of villages return from fields, and children carry their books home from school or herd livestock back to their pens. They return to women – for it is usually still women – at work cooking the family's evening meal. Slowly, the village air fills with the odours of burning wood, dung, coal, straw, charcoal or kerosene. But the smoke is coming from inside the homes, where the women work and their families gather in air growing thick with choking fumes.

Almost half the world's households eat food cooked on open fires and inefficient stoves. These cooking places are death traps for some of the world's most vulnerable people. Levels of smoke and gases such as carbon monoxide are often hundreds of times higher than those that would be tolerated in factories. Women toiling in these fumes for several hours each day suffer, often along with the young children gathered around them.

Wheezing and itchy eyes are the superficial signs of the health effects. But the constant assault on lungs creates a silent epidemic of disease. Children are left vulnerable to pneumonia, which is the biggest cause of death among under-fives worldwide, and the largest single cause of loss of 'life years' for our species. Their mothers are more vulnerable to TB. Pregnant women risk stillbirth and their babies reduced lung capacity. Older people are plagued by lung cancer, high blood pressure and heart disease. The death toll from this indoor pollution is estimated at 3.8 million a year, including more than half a million children under the age of five.

The solution is, on the face of it, simple: the distribution of better cooking stoves that burn more of the fuel and create less smoke and fewer fumes. There is no shortage of designs. Dozens of NGOs have worked on the problem. Many improved stoves can cut fuel needs by half and reduce emissions by even more. They can cost as little as three dollars each. The problem is scaling up manufacture, and getting the clean stoves to the half a billion households who need them, in some of the poorest and most remote parts of the world – and making their use affordable and (sometimes) socially acceptable.

A simple stove has a combustion chamber where the fuel burns, usually over a grate through which air rises, feeding the flames. The hot gases given off by the fire heat a cooking plate on which the pots and pans rest. Some stoves also have a simple chimney through which some of the smoke escapes outside. But most stoves burn their fuel inefficiently, and much of the heat is lost. Incomplete combustion maximises the escaping smoke and unburnt gases.

Improved designs can reduce these hazards. Better air circulation makes burning more efficient, so less smoke and gas is produced. More efficient ways of removing the smoke and gas improve the air inside the house. Insulating the stove can reduce heat waste. A well-placed chimney will also draw air through the stove, improving burning. More advanced designs have fans to remove smoke.

Despite many initiatives, few efficient stoves are produced in the numbers needed to meet the scale of the challenge, says the London-based Ashden Trust, which has been giving awards for good local designs for the past decade. Either they are too expensive for the poorest people who need them the most, or they are too cheap to make them profitable for manufacturers. Distribution is also difficult. And even when the stoves make it to the villages, take-up can be poor.

Nobody has quite pinned down why. According to a study by Grant Miller of Stanford University School of

Medicine, some women say they rather like the smoke from their old stoves, because it keeps away the malaria-carrying mosquitoes that bite their children. Others say they fear new stoves might mean their food tastes different, angering their husbands. Or the designs on offer simply do not match the particular local cooking methods, whether that is standing up to make tortillas in Mexico, squatting to make dhal in India, or tending African bucket stoves to make bread.

This is a problem crying out for an effective solution. It shouldn't be hard, should it? Activists seem divided about whether it needs a global effort, like the drive a decade ago to bring safe drinking water to the world, or whether the solutions need to be local. Meanwhile, the death toll continues to rise.

Fix Our Climate

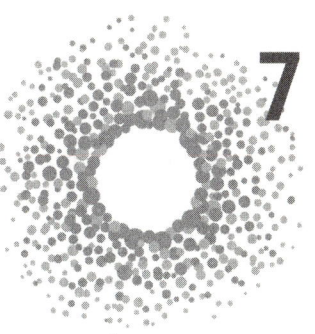

If you don't believe that climate change is happening now, just walk out of your door. You'll see birds and butterflies migrating at times different from when they used to. The cherry blossoms in Washington DC came out two to three weeks earlier this year than they have ever done before. Scientists know that trees are moving to higher altitudes because they need more water and cooler temperatures. This is not normal. It is nature trying desperately to adapt.

If that seems perhaps interesting but inconsequential then watch the news for a couple of months. The scale and geographical spread of fires, floods and other extreme weather events are a reflection of unprecedented disruptions in the climatic system.

But it's what we don't see in the news that angers and upsets me the most: the poorest and most vulnerable people on our planet whose lives are getting even harder as climate change bites. People who have done the least to cause climate change are all too often the ones who have their lives impacted the most. This rarely gets reported by the media, but it is that injustice which has motivated me every day for the past thirty years to continue working to address climate change.

Frankly, I often feel despair about our delay to action. But I know I cannot stay in that dark box. If I am to contribute to the solution I have to move through to the light. So, I think of myself as a stubborn optimist, powered by the gritty conviction that, despite all the setbacks, together we can harness the necessary courage and ingenuity to face the challenge. And there are two major reasons I have for stubborn optimism.

Firstly, climate change is clearly a human problem, here and now; it's not one facing some far imagined future generation, but those alive today. That realisation brings the urgency of the matter into focus. In 2001 if I told someone we had just thirty years to reverse climate change they often nodded politely and moved on. Now if I tell them we have at most a decade to halve global emissions, they often look very scared. Perhaps they think about their children or grandchildren or they worry about the business they run or country they lead, but the penny has dropped, they know it's not going to be someone else's problem anymore – it's ours. Ironically that sense of urgency means we have a chance – there's lots to lose by not acting and vast amounts to gain by moving fast. This motivation works well for a human brain evolved to be *capable of the most extraordinary long-term planning ever seen in nature, but emotionally wired to respond primarily to short-term threats and opportunities.*

Secondly, I have seen personally what happens when political leadership enacts effective incentives. My own country of Costa Rica is only the size of West Virginia but contains 5 per cent of the world's bio-diversity. From the 1970s onward national parks started to be established and were then expanded. Today one quarter of our land territory is protected national parks. But even with the best attitude towards nature, people have to make a living, and outside of these national parks mass deforestation soared as trees were cut down for agriculture or cattle ranching or to sell the wood itself. Two decades ago, the country's forest cover had diminished to an all-time low of 29 percent.

It is not realistic to expect people to stop what they need to do in order to survive only because it's in the long-term interest of biodiversity . . . or the climate. People who live in rural areas need to live off the land, so whether that is cutting wood, raising cattle or farming, that is their livelihood, and change can only occur if a viable alternative is offered. So, we enacted legislation to financially incentivise people to reforest their land – the first scheme of its kind in the world. We offered payment for protecting and expanding forests – effectively we paid people to grow carbon sinks! Today, 52 per cent of the country is forested again.

Costa Rica has benefited from its reforestation over the past few decades. It protects our aquifers, provides income from tourism, reduces flooding, and provides the surrounding area with cooler, moist temperatures that helps both wild plants and farms to thrive and makes them more resilient to the coming changes in the climate. But the world also benefits from this reforestation as the Costa Rican forest is now one of the most effective carbon sinks on the planet. We have learned that well thought out, long-term incentives work. We have also learned that ensuring local communities have a stake in the solutions is the only way to effect change at the scale and speed we need. Finally, we have learned that nature can be our ally. It can recover and draw down vast amounts of carbon, which alongside ambitious plans to reduce emissions at source, could be enough to get us back to a safe, stable climate.

I know that change is possible.

We have the opportunity now to be able to look back in 2030 and say that this was when humanity woke up, stood up and addressed the greatest challenge that we had ever faced, forging a safer, healthier planet. That is the future that I want, and that is the future that I invite you to.

Christiana Figueres, 2021

For a moment, as the plane heads north from London up the east coast of England, the flickers of white below could be seagulls. But no. What passengers can see below, all the way to Scotland, is a technological revolution. Once, there were just waves lapping on to sandy beaches, and the occasional pleasure boat. But now the waters just offshore are peppered with giant wind turbines on some of the world's largest wind farms.

From the London Array in the Thames estuary, past Gunfleet and Sheringham Sands, Humber and Hornsea, currently the world's largest, to the Inch Cape Wind Farm off the Angus coast in Scotland, the farms contain more than 2,000 turbines, some up to 300 metres tall and generating as much as 10 megawatts each. Some days, they produce most of Britain's electricity.

In the past decade the shallow waters of the North Sea have become the world's proving ground for these offshore giants. They may be more costly to build than wind turbines on land. But they benefit from the stronger and more persistent winds that buffet the North Sea for most of the year, and they avoid the problems of finding suitable sites on land. They can be erected in shallow waters to deliver, when working in harness, as much green electricity as a big coal-fired power station.

All of the twenty biggest offshore wind farms in the world are in Europe, with half in Britain alone, and the rest across the North Sea off the Netherlands and German coasts. Jobs manufacturing the giant turbines have replaced jobs lost with the decline in construction of platforms to extract oil and gas from beneath the North Sea. If passengers on flights over the turbines look inland, they may also spot, glinting in the sun, many of the 1 million rooftops and thousands of farm fields across Britain now covered in solar panels. They too are generating clean, green electricity, all part of a new industrial revolution happening in real time, at scale and at pace. The climate crisis is upon us. But so is a solution.

The media cover climate change almost every day, whether bringing news of worsening scientific predictions or analysing the politics of taking action in time to reach the target agreed in 2015 to keep global warming to within 1.5 degrees of pre-industrial levels. That means ending our emissions of the gases that accumulate in the atmosphere causing the warming. In particular it means rapidly reducing emissions of carbon dioxide – a gas released by burning fossil fuels and deforestation.

Once in the atmosphere, carbon dioxide sticks around for centuries. So to have a good chance of keeping warming below 1.5 degrees, we cannot simply cut those emissions. We need to halve them by 2030, and end them approximately a decade later. After that, any gas we add must be balanced by projects to remove from the air more carbon dioxide than nature currently manages.

We also need to act against other human-made warming gases, notably methane. It doesn't last as long in the air as carbon dioxide, but while it is around it packs a punch.

Achieving such targets will be a big ask. Global emissions have risen 60 per cent since 1992, the year the world's leaders met at the Earth Summit in Rio de Janeiro and agreed to prevent 'dangerous' climate change. Emissions may at last be stabilising, but the downturn has not yet begun. Still, the prospects have improved greatly in the past five years. Because, even though the politics continues to move slowly – and we consumers are only just beginning to change our lifestyles to reduce our carbon footprint – technological innovation is gearing up. And whether it is tasty meat-free meals or electric vehicles or solar panels on roofs, we are often keen to take them up.

From giant solar farms and Britain's offshore wind turbines, from smart farming to new industrial processes, from electric cars to blueprints for hydrogen-powered planes, from proposals for geo-engineering to restoring nature's great carbon stores, and from stopping cows belching methane to plugging leaky pipelines: we are giving ourselves a fighting chance to reach the target. If we will only grasp the opportunity.

Britain is an unlikely trailblazer. It arguably started the climate crisis more than two centuries ago, when it developed the steam engine and began powering the first giant factories with coal – thus triggering the industrial revolution. But while the UK exports emissions by having other nations produce much of its food and goods, progress on renewable energy at home has been impressive.

Half a century ago, it still generated most of its electricity by burning coal in some of the world's biggest power plants. Yet now it is a pioneer in the counter-revolution. By 2025 the last British coal-fired power station will have shut. Some of these giant edifices have been turned into art galleries or luxury hotels. The largest, Drax power station in Yorkshire – whose twelve cooling towers dominate the horizon for miles – now burns a renewable fuel, wood pellets, although they are brought in barges from the southern United States.

The United Kingdom is one of thirty-two countries to have achieved what was once thought impossible – powering ahead with economic growth while dramatically cutting emissions. Others include the United States, France, Germany and Japan. Since 2005 British emissions are 33 per cent down, while gross domestic product (GDP) is 20 per cent up. Although as we have already noted this doesn't include emissions 'embedded' in British imported goods.

In most cases, this decoupling of the link between economic growth and carbon dioxide emissions persists even after taking account of the emissions produced during the making of goods imported from China and elsewhere. In any case, China is going in the right direction too. It has halved the amount of carbon emitted for every dollar of its GDP since 2005. While China's rapid economic growth means its emissions continue to rise, this shows that economies in developing countries can decouple from emissions.

This decoupling has happened through a combination of widespread switching to ever-cheaper low-carbon energy sources, such as wind and solar power, and the more efficient use of energy in industrial processes, home appliances, heating systems and much else. It is a start. But we need massively more of both.

Low-carbon forms of energy are growing in popularity as costs of investing in them come down, often to below those of new fossil fuel plants. Countries are finding many different ways of generating electricity that break the hegemony of coal, oil and natural gas.

In Iceland, geothermal energy generates a quarter of the country's electricity and heats 90 per cent of all homes. In South Korea, they have the world's largest tidal power station, fuelling Siheung, a city of half a million people. Britain loves wind; Norway binges on hydroelectricity; many countries get a lot of heat from burning wood and other renewable biomass products. In Morocco, they are installing half a million mirrors in the Sahara desert that will direct solar radiation to heat sodium salt and run turbines. The aim is to export this solar thermal energy to Europe.

Coal-dependent, smog-afflicted India – regarded as a near-pariah state as recently as 2015 at the Paris climate negotiations – is moving fast towards solar power. Even oil superpower Saudi Arabia says it wants half of its power to come from renewables by 2030, with the emphasis, unsurprisingly for a desert state, on solar energy. Coal plants are still being built, but investment in new electricity-generating infrastructure is now predominantly for solar and wind power.

The solar story in particular is phenomenal. Less than half a century ago, it was so expensive that solar panels

were used only in spacecraft. But its price has fallen dramatically – by more than 90 per cent in just the past fifteen
years, making it often cheaper than coal. Solar power is
on the verge of becoming, as one academic put it, 'the
least costly source of electricity production in the history
of mankind'.

Countries are at vastly different stages of weaning themselves off fossil fuels and cutting their emissions. But the
potential to decarbonise is there. One study found that the
amount of carbon dioxide emitted for every unit of electricity generated ranges from 3 grams for every kilowatt
hour in Norway, where hydroelectricity dominates, to 100
grams in nuclear-powered France, 600 grams in much of
the United States, 900 grams in India and more than 1,000
grams in parts of energy-inefficient coal-burning China.

The direction of travel is clear. Recent trends suggest
that the world has probably hit 'peak coal'. As electric cars
take off, many predict it will soon reach 'peak oil' – even
the oil giant BP agreed, in its regular energy outlook.

There will be downsides to the new energy technologies.
Big solar farms are land-hungry. Wind turbines are sometimes noisy and can be bad for birdlife. For instance, the
giant wind farms being erected off east Scotland could
disrupt the lives of thousands of seabirds. Scotland is
home to 45 per cent of Europe's breeding seabirds, living
in staggering densities on cliff faces and volcanic islands.
Bass Rock, near North Berwick, is home to 150,000
gannets, many of which feed on sand eels they find on
submerged sandbanks, which are being targeted for wind
farms. 'Nowhere in Europe are there such internationally

important colonies so close to such major wind farms,' says Tom Brock, former boss of the Scottish Seabird Centre, which is based nearby, in an interview with one of the authors for *Yale Environment360*.

But advances in the efficiency of energy transmission may help reduce such conflicts. They will allow solar farms to be located deep in empty deserts and wind turbines to be erected further offshore. Making such intermittent energy sources our main means of generating electricity will require massive energy storage, to enable the lights to come on when the sun is down and the winds don't blow. But the cost of battery storage has fallen by 90 per cent in a decade, with more to come as the technology develops.

Right now, the switch to a solar- and wind-powered world seems all but unstoppable. And not just in rich countries. For the billion-plus people round the world living in remote villages where electricity grids have not yet reached, renewables are probably the best way of plugging in, through innovations such as local microgrids, often powered by village solar panels.

• • •

Plugging into electricity for the first time is a big deal. Ask Peter Okoth. Until late last year he struggled to make a go of his bar on the main street in Entasopia. The dusty one-street town in Kenya's Rift Valley is 50 kilometres from the nearest grid power line. A solar panel on his shed roof powered one light but no more. Then, he hooked up to a new solar-powered microgrid installed by a Kenyan start-up company to serve the town's homes and businesses. Now Peter has eleven light bulbs, and enough power to run a TV and a sound system for his customers. Seventy people show up some evenings to watch, listen and buy his food and drink. His profits will soon buy a refrigerator to keep the beer cold, and a big screen to show satellite sports channels. 'We will be staying open till midnight,' he says, in an interview with one of the authors for *Yale Environment360*. He has just bought construction materials for ten guest rooms. 'When you next come, you must stay here.'

Most rural settlements in Kenya are dark at night. Only a third of the East African country's residents have access to the national power grid. So harvesting the sun locally makes obvious sense in hundreds of communities like Entasopia. Hundred-dollar solar panels for installation on individual home roofs have been on sale for years. But the meagre five watts that most such systems provide is only enough for a couple of LED lamps and a mobile phone charging point. And the batteries constantly need replacing. The country is full of discarded solar cells, defunct batteries and disappointed customers.

But now, larger village solar power units linked by cable to dozens of houses and business, and with sophisticated microchip control systems to manage supply, are starting to transform lives. For a ten-dollar installation fee, Peter and his neighbours buy a share in a thousand times as much power as before. Village homes are filling with household appliances like refrigerators and washing machines, and businesses on the main street are powering everything from welding equipment and fuel pumps to hair dryers and karaoke machines.

Microgrids can rely on local river hydropower, wind and biomass, as well as the power of the sun. There are no official statistics on how many there are, how many people they benefit, or what their total output is. But they are, says Daniel Kammen of the University of California, Berkeley, in an interview with one of the authors for *Yale Environment360*, 'a true hotbed of innovation popping up all over the world'. And the energy generation does not lead to a gram of extra carbon dioxide reaching the atmosphere.

• • •

Electricity is taking over the world. Soon, almost no home or community will be too remote. Road vehicles too seem set to become electrified. But ships and planes cannot easily plug into the grid. And some industrial processes require burning coal and other fossil fuels as part of their chemical transformations. These hard-to-fix industries and transportation systems are responsible for more than a fifth of global carbon dioxide emissions. So, if we need to get to zero emissions, what should be done to fix them?

Planes are generally built to minimise fuel use. They are made of aluminium, controlled by smart systems, and fly in the thin air 10 kilometres up. Their carbon dioxide emissions per passenger kilometre have halved since 1990. But as demand soars so they remain among the fastest-growing sources of the gas in the air, even as airlines have promised to cease further increases. They will be looking

209

for further savings. One source of waste is the fuel they burn maintaining holding patterns while waiting to land and while taxiing on the ground. They also take surprisingly circuitous routes to their destinations, minimising travel over oceans and skirting round unfriendly countries, as anyone watching the tracking maps on an international flight will have noticed. Clearly, however, fixing such things will do little to help the world towards zero emissions.

Beyond emitting carbon dioxide, aircraft contribute to warming through other emissions – notably water vapour. It freezes in the cold upper air to make condensation trails of ice crystals that trap heat in the lower atmosphere. By some accounts, these contrails more than double the contribution of aircraft to warming, from 2 to 5 per cent. The calculation is not exact. The effect of contrails wears off within minutes as the ice crystals dissipate, whereas the carbon dioxide sticks around for centuries. So during the height of the Covid pandemic, when flights were much reduced, the warming effect of contrails dropped, while the carbon dioxide from past flights remained.

Still, cutting contrails would keep us cooler. One solution is to fly lower, where the air is warmer and water vapour is less likely to freeze. Studies suggest that requiring planes to fly at heights where they don't generate contrails could reduce warming from flights by about a tenth even when also taking fuel use into account. Currently, however, there are no international agreements to encourage airlines to do this.

The real challenges for decarbonising aircraft are to limit flights – why fly to a business meeting when the pandemic has shown how video conferencing is so convenient? – and to find new fuels for those that continue. On the latter, various ideas have been proposed. Aircraft could be one area where burning biofuels could be helpful. Some see our skies filled in future with drone-style self-driving electric heli-planes, operating like taxis. They are already being lined up for short sight-seeing flights over islands in the Pearl River estuary in southern China.

But the smart money is currently on hydrogen, which has the added advantage of being three times as energy dense as kerosene or most biofuels, and so lighter to carry on flights. Airbus, the world's second largest plane maker, is putting its money on hydrogen. It has three different zero-emission 'concept' planes. California start-up ZeroAvia has a six-seater research plane already running on hydrogen. It took off from Britain's Cranfield Airport for the first time in 2020. Could this be the Tesla of the skies?

The race is also on to decarbonise the 50,000-plus tankers, freighters, container vessels and ferries that make up the world's shipping fleet. They could try slowing down. When world trade dropped after the 2008 financial crisis, Maersk, the world's largest shipper of containers, ordered its captains to sail more slowly so they spent less time in port. It discovered this go-slow cut fuel use 30 per cent. Longer-term, new ships could be built of lighter aluminium, or with a more slender design to slide through the water more easily. But what they really need is new means of propulsion.

Electric engines already operate on some short ferry journeys, but the sheer weight and space taken up by batteries makes electric propulsion unviable on ocean-going ships for now. Unless they have on-board sources of green power. So, could we be headed back to the days of sailing ships, when wind power ruled the waves? It is not so far-fetched. The *Viking Grace*, a cruise ship now operating between Finland and Sweden, boasts a 'rotor sail', a large spinning cylinder amidships that catches the wind. It reduces fuel needs by up to a fifth.

The Japanese designed *Super Eco Ship 2030* would have 4,000 square metres of sails to catch the wind and cut emissions by 70 per cent. Going one better, a Scandinavian shipping line says it has a design that would transport up to 10,000 cars (electric, we trust), powered by electricity generated on board from a mixture of wind, solar and wave power. The vessel could be afloat by 2025.

• • •

Some say we should forget electricity for transportation altogether. That the rush to electric cars is a false dawn. They see hydrogen as the new wonder fuel, powering trucks, ships and planes, and making everything from cement to concrete, even our food. As mentioned, hydrogen contains more energy for every tonne than any fossil fuel. And when you burn it, all it emits is water. It is also a convenient way of storing energy without using batteries. So, the argument goes, we should chart a course for a 'hydrogen economy' to replace our existing fossil fuel economy and save the climate from carbon emissions.

Hydrogen power is not a new idea. The first hydrogen-powered engine was working more than two centuries ago, in 1807. Scientists proposed making hydrogen from water to replace coal as early as the 1860s. In 1922 British biologist Jack Haldane mused that 'ultimately we shall have to tap . . . the wind and the sunlight', and that the world would be 'covered with rows of metallic windmills working electric motors . . . during windy weather the surplus power will be used for the electrolytic decomposition of water into oxygen and hydrogen'. But coal and oil were always cheaper. And the Hindenburg disaster in 1937, when a hydrogen-filled airship exploded, gave it a largely undeserved reputation as unsafe.

But maybe now hydrogen's time has come. A century after Haldane's prophecy, Japan in 2021 planned a 'hydrogen Olympics'. The Olympic torch travelling through the country burned hydrogen. The Olympic village was hooked up to run on the products of a solar-powered hydrogen manufacturing plant built inside the exclusion zone created after the Fukushima nuclear accident a decade ago. Hydrogen buses and hydrogen VIP cars were on hand. Hydrogen may, as one media outlet suggested, be about to morph 'from a niche power source – used in zeppelins, rockets and nuclear weapons – into big business'.

There is one problem. Hydrogen has to be manufactured, and that takes a lot of electricity. So, hydrogen is only as climate-friendly as the energy used to produce it. Engineers distinguish between blue hydrogen, which is made from fossil fuels and has a big carbon footprint unless the carbon dioxide is captured, and green hydrogen, which is made by splitting water using renewable energy, and need have no carbon footprint at all. In time, countries with abundant renewable energy could become the Saudi Arabias of hydrogen: Canada, with its hydroelectricity, Iceland, with its geothermal energy, or Morocco, with its solar power from the Sahara desert, for instance.

But Saudi Arabia has its own plans for green hydrogen. It plans to cover a patch of desert the size of Belgium – near its proposed eco-city of Neom on the Red Sea – with a vast estate of solar panels and wind turbines. By 2025 they would be powering the world's biggest hydrogen factory. Neighbouring Oman has plans to go even bigger. Australia says it will turn an area of desert more than twice the size of Luxembourg into a green hydrogen production facility, with 10 million solar panels and 1,500 wind turbines. It may send the hydrogen to Japan, where Kawasaki has built the first ship to carry liquified hydrogen, aimed at tapping Australian production. Danish wind-power company Orsted has signed up with shipping conglomerate Maersk and airline SAS to harness offshore wind from the North Sea to produce green hydrogen for ships and planes.

Will the hydrogen economy happen? Maybe not. A major downside for green hydrogen is that it is not a particularly efficient fuel. Every step of manufacturing, transporting and use loses energy. Hydrogen vehicles, home heating and other proposed uses would be only a third as efficient as electric equivalents. Many say that should rule out hydrogen for these uses. But where electricity is not viable – for aircraft, ships or big energy-guzzling industries that engineers despair of making carbon-neutral – then hydrogen could be the thing. Joe Biden's energy secretary Jennifer Granholm says 'clean hydrogen is a game changer' because it 'will help decarbonise high-polluting heavy-duty and industrial sectors'.

These include three of today's biggest fossil fuel burners – steel, cement and chemical fertilisers – which cannot be easily converted to electricity because burning fossil fuels is part of the process itself. Between them, these carbon-spewing monsters are responsible for up to 15 per cent of global carbon dioxide emissions. Many experts see hydrogen as the only available route to curb them.

Take steel. Its production starts with iron ore being smelted by burning coal in a blast furnace to make iron. Besides heating the furnace, the coal is the 'reducing agent', removing oxygen to leave behind pure iron. Unfortunately, that also makes carbon dioxide. However, coal can be replaced as the reducing agent by hydrogen. That could become the norm. Swedish iron mining and steel-manufacturing companies have a plan to replace coal in the furnace with green hydrogen to make 'fossil-free' steel. Australia could do the same: it has both big iron mines and plentiful solar energy to make hydrogen.

Nitrogen fertiliser production could also benefit from a green hydrogen makeover. It is currently made from the century-old Haber-Bosch process, which produces 2 tonnes of carbon dioxide emissions for every tonne of fertiliser. This is partly because the main process involves reacting nitrogen from the air with a concentrated stream of hydrogen that is currently made from fossil fuels. A Japanese engineering company, JGC, has a trial plant using solar energy to make the green hydrogen. That would reduce emissions a lot.

Finally, cement manufacture also has a carbon dioxide

problem. When the fuel in the kiln is coal, about a tonne of carbon dioxide is produced for every tonne of cement. It too requires lots of energy, and when limestone (calcium carbonate) is converted in the kiln to calcium oxide, the main by-product is carbon dioxide. Green hydrogen could replace the fossil fuel, but that would only reduce emissions by a third. Something more radical is needed.

Enter Australian technologist John Harrison. He has demonstrated a way to make cement that soaks up carbon dioxide. All we need to do, he says in an interview with one of the authors for *New Scientist*, is replace limestone in cement with another rock, magnesium carbonate. This alternative makes cement that is just as good, he says. But it roasts at lower temperatures, halving energy need. Even better, the resulting cement will reabsorb from the air almost as much carbon dioxide as was lost originally.

This reabsorption process is called carbonation. Regular cement does it too. But it soon shuts down. Whereas carbonation in magnesium cement just keeps on going, for as long as it is in contact with the air, Harrison says. A bridge, road or tower block made from his 'eco-cement' could be performing much the same eco-function as a tree – gradually absorbing carbon dioxide throughout its life. If this can be done at the scale Harrison believes while maintaining the efficacy of the concrete then instead of being one of the world's biggest sources of carbon dioxide emissions, the cement industry could be close to carbon-neutral. Our 'concrete jungles' would be as green as the Amazon rainforest.

Magnesium carbonate is not as abundant as limestone.

But it is found as the mineral magnetite and in dolomite rocks. So why isn't the cement industry taking advantage? Industry insiders say there have been no serious efforts to try out Harrison's proposal on a large scale. Fred Glasser of the University of Aberdeen's chemistry department says the real problem is that the building materials industry is 'intensely conservative'. 'It prefers what it knows and what is cheap,' he says in an interview with one of the authors for *New Scientist*. Perhaps the first company brave enough to take the plunge could corner a massive new market.

· · ·

Just as important as switching the world to low-carbon energy is reducing the amount of energy we need. That can be done through big improvements in energy efficiency everywhere from industrial processes to electricity for washing machines and bedside lamps, from air conditioning to space heating.

Energy efficiency begins at home. In cold weather, our notoriously leaky homes need much better insulation to hold heat in. In a country like Sweden, lagging the loft, double- or even triple-glazing windows and insulting cavity walls can halve heating bills and with them the emissions from heating. In hot climates, we need a revolution even more urgently, as the world warms.

Any visitor to Singapore will know the story. Outside it is blazing hot. Inside tower blocks and offices, the air conditioning it turned up so high, people are shivering in an air-con breeze set at 20 degrees or lower. Jackets are put on to keep warm. All that cooling inside adds to the heating outside, cranking up the 'urban heat island' effect by as much as seven degrees. Singapore uses more energy for air conditioning than probably any other city in the world. The city says it can only meet its Paris climate targets if its buildings cut energy demand. Don't run air conditioners at full blast: 25 or even 28 degrees is fine. Make shirtsleeves the dress code. Oh, and make air conditioners more energy-efficient. Thirty per cent gains would be easy, says Lee Poh Seng of the National University of Singapore, in the newsletter *Eco-Business*.

Passive cooling should be a must, too. White paint helps. White roofs keep hot buildings cool in the sun by reflecting up to 90 per cent of solar energy. That is no secret. Buildings around the Mediterranean have been built of white material or painted white for thousands of years. It is a lesson worth heeding as the world warms.

But there could be even more to come from the lab. White is white, you might think. But Xiulin Ruan at Purdue University in the United States has come up with a super-white paint. It replaces white gypsum or titanium dioxide pigment with even whiter barium sulphate. This pigment raises the reflectivity of white paint to 98 per cent, reducing temperatures inside by up to five degrees. Paint the roof of your house with this paint, Ruan told fellow Americans in the press release for the study, and 'you could get a cooling power . . . more powerful than the central air conditioners used by most houses'. The paint could be in stores within two years.

Air conditioning is a big user of energy in some industries, too. Some of them have no need to be in hot climates at all, and ought to be moved somewhere cooler. Take digital data centres, the humming hubs of our modern interconnected world. The world's digital revolution has been a boon for reducing energy use in many ways, allowing handheld phones to do things that once required myriad different devices, and enabling us to work from home and communicate by video conferencing. But there is a dark side to that revolution: the data centres that drive them.

These factories of the digital age sit in giant windowless warehouses across the globe, consuming vast amounts of electricity. The biggest data centres consume as much power as a city of a million people. Google estimates that a typical internet search requires as much energy as running a 60-watt light bulb for 17 seconds. But searches pale into insignificance compared to streaming services. IT conglomerate Cisco reckons that video streaming makes up to 80 per cent of internet traffic. Collectively, data centres emit as much carbon dioxide as the global airline industry.

Some tech companies recognise that is a problem. They are switching to renewable energy, which is good. But in internet hubs such as India and China, most of the energy running data centres still comes from burning coal. That must change. But, whatever the fuel, there is also an urgent need to reduce energy use.

A lot of the energy expended in data centres is not doing computing, it is keeping the whirring processors cool. Not least because, for historical reasons, many are in hot deserts like Nevada or southern California. The obvious solution is to move the centres to cooler climes. Google and Microsoft have been building hubs in Finland and Iceland. 'It's all natural cooling here,' says chief technology officer Tate Cantrell at Verne Global, a data hub in Iceland, in an interview with one of the authors for *Yale Environment360*. 'Our chiller is an open window.' A major data centre in Norway is cooled by freezing water from a fjord.

The next step to sustainable data centres will be to find uses for the heat still given off by warehouses full of processors. One data centre in Stockholm heats 10,000 nearby homes. Project mastermind Fredrik Jansson says in an interview with one of the authors for *Yale Environment360*: 'When my son is on his iPad at home in Stockholm, I can tell him his digital activities end up helping to heat the radiator in his room.'

• • •

The pace of technological change in the past five years is staggering. It no longer looks ruinously expensive to fight climate change. The trillion-dollar price tags have disappeared. As a result, many governments and giant corporations promise to achieve net zero emissions long before the global deadline of 2050. And the call from climate scientists for the world to halve emissions by 2030 no longer looks outlandish. The European Union promises a 55 per cent cut by 2030. Amazon, which already claims to be 'the largest corporate purchaser of renewable energy', says its operations will be fully powered by renewables by 2025. To do that, it is paying up front for, among other things, a new Scottish offshore wind farm.

Even the energy industry is buying into the idea. In 2021 the IEA proposed a route to net zero that effectively pulled the plug on big fossil fuel as of now. Its plan ended all oil and gas exploration, and all further approval for new coal-fired power stations, with immediate effect. Cripplingly expensive? No, said its director. The result would be a net economic benefit to the world.

Such goals won't necessarily be achieved just because an industry body makes a plan. In the aftermath of the IEA report, the United Kingdom was still licensing new oil and gas fields in the North Sea, China continued to build new coal power plants, and oil companies were still drilling. But the mood is shifting fast. It could be that a tipping point has been reached in the fossil fuel business.

Days after the IEA report, America's Biden administration halted plans to give leases for oil exploration in Alaska. Then the government of coal-mining giant Indonesia announced it would approve no more coal-fired power stations. And finally, Germany's Supreme Court told the federal government in Berlin to toughen its targets for cutting emissions by 2030, because they did not do enough to protect future generations from the consequences of climate change. The government promised to do what it was told.

Aid agencies have been shifting too. The Japanese government promised at a G7 summit to join the other six nations of the club in banning all future aid for coal-fired power plants. The Asian Development Bank said that it would go one better and start paying for countries to shut their coal plants early. That is a vital step in a continent where, according to the IEA, the average age of coal plants is just eleven years, meaning most could still be operating in 2050.

Most dramatically, the pressure was getting to big oil companies. When Shell announced that its own road map to zero carbon emissions would include an increase of 20 per cent in natural gas production in 2030, there was a shareholder revolt, led by Britain's biggest fund manager, Legal & General. Then a Dutch judge held the company liable for helping cause dangerous climate change, and ruled that it must cut its emissions by 45 per cent by 2030 to comply with the Paris Agreement – the first time a judge had ordered a big polluting company to comply. He said there was 'a broad international consensus about the need for non-state action, because states cannot tackle the climate issue on their own'. Commentators said it set a precedent for other big oil companies.

By chance, on the same day as the Dutch court ruling, hedge fund climate activists in the United States won the support of mainstream investors to secure two places on the board of ExxonMobil, and shareholders at Chevron voted to force management to start cutting emissions. 'It may be the most cataclysmic day so far for the fossil fuel industry,' said veteran climate activist Bill McKibben. 'What sounded radical a decade ago now sounds like the law.'

• • •

So are we on a home run to zero emissions? Is the climate problem solved? Not so fast.

Even if all the promises and plans and road maps to net zero are fulfilled, the world may warm faster than scientists predict. Or critical tipping points in the climate system may arrive even below 1.5 degrees of warming. Nobody can be sure. So what then? If things get too bad, we may need a Plan B to pull carbon rapidly from the air, and haul us back from the brink.

Enter the notion of geo-engineering, defined by scientists as 'the deliberate large-scale intervention in the Earth's natural systems to counteract climate change'. There are two main approaches to this. Both sound like science fiction.

One is to cool the planet by shading it from solar radiation. As long ago as 1997, Edward Teller, inventor of the hydrogen bomb, proposed doing this by putting giant mirrors into space. Other ideas include covering the oceans in white polystyrene balls, pumping salt particles from the ocean into clouds to make them brighter, and mimicking the cooling effect of volcanic eruptions by pouring light-scattering sulphate particles from planes into the stratosphere.

This last proposal could be the quickest, most effective and least costly. Even so, the planes would have to keep flying day after day, and year after year, spraying ever-larger quantities. If they stopped, the particles would be washed out by the rain within weeks, and the world would resume warming with redoubled force. And even then, the suppression of solar radiation could create massive changes in weather systems, maybe shutting down the Asian monsoon on which 2 billion people depend for their food crops. The continued accumulation of carbon dioxide in the atmosphere would have many other effects that will not be neutralised, such as acidifying the oceans.

A better approach is to directly remove carbon dioxide from the atmosphere – so-called negative emissions. We could mimic nature with millions of artificial trees covered in chemicals that absorb carbon dioxide, though they could take up as much land as planting real trees, and could be rather more expensive. Quicker and less land-hungry would be building machines that actively suck up air to strip out the carbon dioxide. They would require lots of energy, and it would need to be generated without emitting carbon dioxide or the gains would be lost.

This idea was first suggested in 1995 by Klaus Lackner, now at Arizona State University. And while direct air capture has long been seen as outlandish and ludicrously expensive, potentially costing hundreds of trillions of dollars, like the price of wind turbines and solar panels, the predicted bill keeps coming down. 'There's been very rapid progress,' says Stephen Pacala of Princeton University in an interview with *Yale Environment 360*. 'So much so that knowledgeable people, who are not starry eyed but just hard headed, believe there is a high probability that a research effort within ten years would produce direct air capture at a hundred dollars a tonne.' That makes it doable, he says.

It still sounds crazy – and remains more costly than fixing the problem now, before things get so bad. Critics say even researching such technologies is dangerous. By suggesting any kind of easy fix for global warming, it encourages delay in ending our addiction to fossil fuels.

But we may need a Plan B, says former British government chief scientist David King, who has long warned ministers of the risks of climate change. He has set up the Centre for Climate Repair at Cambridge University, the first major research centre dedicated to a task that, he says, 'is going to be necessary'.

He is not alone. In the United States, the National Academy of Sciences has launched a study into sunlight reflection technologies. Marcia McNutt, the academy's president, said in the press release: 'We are running out of time to mitigate catastrophic climate change. Some of these interventions . . . may need to be considered in future.' China, meanwhile, is looking at how solar shading might slow the rapid melting of Tibetan glaciers.

But is geo-engineering the right Plan B? Many say that if we need ways of dragging carbon dioxide out of the air, harnessing nature will be easier, cheaper and much more environmentally friendly. 'Nature-based solutions' require boosting ecosystems we know to be great carbon absorbers, such as forests, peatlands and 'blue carbon' coral reefs, seagrass meadows, salt marshes and mangrove swamps.

The conservation case is compelling. The World Resources Institute estimates that globally there are 20 million square kilometres of forests degraded by logging or shifting cultivation that could be restored. That is an area twice the size of Canada. Mostly, given the chance, the forests will regrow naturally. We can also block drainage channels on peatlands, protect coral and plant mangroves. Whatever we do, every tonne of carbon in natural ecosystems is one less tonne in the atmosphere.

Climate scientists have two big concerns. One is that climate change itself will undermine all our efforts to bolster carbon-containing ecosystems. A second is that we have to make sure these nature-based solutions must happen as well as cutting emissions, rather than in place of them. UN climate negotiations have agreed generous terms for polluters to 'offset' their continued emissions by investing in natural restoration projects. The idea is to do whatever is cheapest. Which makes sense. But critics say a lot of projects are carbon frauds. They don't absorb much carbon, or the conservation would have happened anyway. Every time such a project happens instead of cutting emissions, it takes us further from a solution, not nearer.

Still, all things being equal, encouraging existing land-holders to hold more carbon on their territories makes obvious sense. Creating and restoring forests, boggy places and wetlands will all help corral the carbon. And farmers, whose soils are among the richest carbon stores of all, can be part of the solution too.

Right now, many farmers are part of the problem. Organic matter in soils worldwide contains an estimated 2 trillion tonnes of carbon – three times as much as is in the atmosphere. But most of the world's agricultural soils are unhealthy, eroding or losing fertility. They are giving carbon up to the air. A quarter of all the carbon emitted by human activity since the industrial revolution has come from soils. More could be coming as climate change takes hold. So if we want a net-zero world, we have to find a way of calling a halt to the process. The good news is we can. Less ploughing of fields and less intensive livestock grazing on pastures are top priorities.

Tilling fields is almost universal among farmers. It helps seeds to germinate and reduces weeds. But it also exposes carbon in soil to oxygen in the air, allowing microbes to convert the carbon into carbon dioxide. So the recent trend to no-till cultivation holds great promise. Typically, it involves replacing the plough with a drill to make holes in the earth to plant seeds. It slashes carbon dioxide emissions by about a third, and by keeping the carbon locked away, also makes soils more fertile. It's a win-win. That is why half of Australian and Latin American farms do it, along with a quarter of US farmers. What about the rest? And where are Europe and Asia in this revolution?

Livestock farms can capture more carbon too. Ask Charles Massy. He is an Australian farmer, raising sheep on 800 hectares of the increasingly arid Monaro plateau, in New South Wales. He knows the land well. 'My family had farmed in this tough dry country for almost a hundred years,' he says. By his own admission, he drove his 800 hectares to the limit, removing any trees, and grazing the grass till there was little left. It was how things had always been done. 'White Australians like me brought the European mechanical mind to farming. I believed we had to fight against nature in order to survive.'

But the result eventually proved disastrous – for him, for his sheep, for the grasslands where they grazed, and ultimately for the global climate. 'This property was a dustbowl in 1980,' he remembers. Drought took hold. 'We lost the grass. We had soil erosion and plagues of grasshoppers. It was a downward spiral to desertification. I was in debt and sold many of my sheep.' But he had a revelation.

Almost all that was left alive, he says, was a single tree. He had never noticed it before, or he might have cut it down. An Aboriginal friend told him it was an ancient kurrajong or bottletree, planted at least 400 years ago. 'Our ancestors brought these trees from far in the north-west,' says Rod Mason of the Ngarigo people. It mattered a lot to them. 'We planted the tree, we helped and prayed for that tree to grow. We got provisions from it; we camped under it; we lived with it.' The tree's extended trunk allowed it to store water and survive droughts. Its seeds made food; its roots could be squeezed for water; the leaves could be fed to livestock; the wood made shields for fighting, and the bark made fibre for mats.

Rod's story got Charles thinking. 'This old tree, and my Aboriginal friend, helped me realise that to make it as a farmer I had to listen to the land,' he says. Instead of fighting nature, he started to listen to it, and harness it. He believed this was the way to restore his land, and he has been proved right.

First, he protected the soils, by preventing his sheep from grazing for more than two days on any piece of land. 'The animals only take away half the grass before they move on, and that stimulates more growth. Their dung is a natural fertiliser, and the sheep's hooves and dung beetles massage it into the soil.' Less intense grazing also allowed native drought-resistant grasses, known locally as kangaroo grasses, to return.

Held together by roots and stirred by all manner of

insects and fungi, his healthy soil stored more water for use during droughts. And when droughts broke, it coped with that too. 'Last July, my neighbour's land was pouring with mud, while on mine the rain was absorbed in the soil, which doesn't wash away.'

Then he began planting trees. Some 60,000 in all. They sheltered his sheep from sun and winds, but also provided habitat for insect-eating birds that prevented grasshopper invasions. Animals came too. 'We've had two new species of wallaby come in,' he says.

'I have spent the last thirty years transforming this property from a dustbowl into a beautiful and commercially successful farm,' says Charles. His methods have made the land resilient to the droughts that increasingly plague Australia. Even in the bad times now, his sheep can keep producing wool. He doesn't run into debt – and nor do the soils. They keep hold of their carbon and their fertility. 'If you increase the ground cover and your grassland, you are putting a lot more carbon into the soil,' he says. So not only is his land more resilient against climate change, it also helps prevent it.

Charles calls what he does regenerative farming. He sees his neighbours copying it. This is the future. But it is also, he says, harking back to traditional Aboriginal culture, to a way of living in this harsh land that survived for 60,000 years precisely because it worked. 'I see regenerating agriculture and indigenous cultures as both the same,' he says. 'I think regenerative farming is one of the great stories of hope for our planet.'

Rod is pleased by the progress of his star white pupil.

Movers and shakers are noticing, too. In 2021 Microsoft bought credits from a nearby Australian cattle ranch, Wilmot Cattle, which says its own careful regenerative grazing systems have added 40,000 tonnes of carbon to its soils in three years. The farm is owned by venture capitalist Alasdair MacLeod. Suddenly regenerative farming is, he says, 'the new normal'. That is good news for Australia, but good news for the planet's climate, too.

· · ·

Capturing carbon is not the only way that livestock farmers can help fight climate change. Their biggest contribution could be to staunch the flow into the atmosphere of the second most important greenhouse gas: methane. Methane has long been in the shade because the world's attention has been focused on carbon dioxide. But it is growing increasingly clear that unless we can fix the methane problem too, we have little chance of keeping warming below 1.5 degrees in the coming decades.

Methane is produced by bacteria working in conditions without oxygen. Some of it is being produced by natural processes today, including those in bogs and the guts of ruminant animals; some is ancient, namely fossil methane long trapped underground. Natural gas is fossil methane, formed millions of years ago and now being released as we burn it for fuel, or flare it from oil fields.

Methane's concentration in the atmosphere is increasing fast thanks both to our releasing of fossil methane from leaking natural gas pipelines and fracking, and to our influence on current production, including the proliferation of methane-generating rice paddies, landfills and livestock.

Once in the atmosphere, the gas lasts for about a decade – compared to centuries for carbon dioxide. So it doesn't stick around and accumulate like carbon dioxide. But during the time it remains, it packs a punch. The warming effect of every molecule of methane over the decade it is in the air is eighty-four times greater than that of a molecule of carbon dioxide. This means that while curbing emissions of carbon dioxide will only stop temperatures rising further, curbing methane can cause real temperature cuts within a few years. That is why Inger Andersen, boss of the United Nations Environment Programme, says in the press release for a UNEP report that 'cutting methane is the strongest lever we have to slow climate change over the next twenty-five years.'

In the oil and gas industries, plugging leaks in pipelines and ending the flaring of gas from oil wells that result in CO_2 production would make a huge difference. So would reducing fracking. Tapping methane from rotting vegetation leaking out of landfills is smart; you can run power stations on the stuff. But the biggest win is to be found in curtailing the emissions of flatulent cattle.

Cattle are responsible for a third of man-made methane emissions. On average, as they chew the cud, cows produce 4 kilograms of methane for every kilogram of meat protein. Ten times as much as pigs and a hundred times as much as the eggs of chickens. They are, from a climate perspective, methane-producing machines with a sideline in producing meat.

But besides the obvious and essential move to eating less

meat, there is another large potential to mitigate livestock emissions, by changing *their* diets to produce less of the gas. The most spectacular results have come from feeding them seaweed. In the past, cattle in coastal areas have often been fed on seaweed. It was cheap, nutritious and easily available. It still is, and we now know that it cuts their methane output by up to 80 per cent.

That discovery so excited Rob Kinley, an Australian academic, that he set up a company, FutureFeed, to produce a feed supplement containing an extract of the best-performing seaweed. According to his website, he plans to have it on the market in 2023. If just 10 per cent of the world's cattle ate it, he claims, it would have the same impact on our climate as removing 100 million cars from the world's roads.

If he is right, this could go a long way to achieving the international target of cutting methane emissions by 50 per cent in the coming decade. Current estimates are that the cut could reduce global warming in the short term by almost 0.3 degrees. Eliminating emissions by mid-century could deliver a 0.6-degree benefit.

This doesn't mean we don't have to act on carbon dioxide, or work hard to put carbon back where it belongs, in plants, soils and rocks. Far from it. Carbon dioxide levels remain the planet's primary thermostat. But it certainly would help keep down temperatures. 'Methane is by far the top priority short-lived climate pollutant that we need to tackle to keep 1.5 degrees within reach,' says Rick Duke, part of President Joe Biden's climate team, in a UNEP press release.

Build a Waste-Free World

I was born into a community of nomadic cattle
herders in Chad, in the heart of Africa, and so I
have a different feeling than much of the world about
the potential for a waste-free society, because I've
essentially lived in one. For many centuries African
cattle herders, like my family, have lived an almost
waste-free existence. They navigate by the stars and
they depend upon the sun, the rains and the vegetation of the savanna to live. Pastoralists all around the
world make use of everything they find and waste
very little – they leave a very light footprint.

They have learned this way of life from our greatest
teacher, Nature. The natural world finds value in absolutely everything from grass, to dung and even dead
animals. All creatures and plants are part of a perfect,
beautiful system, a world of cycles that teaches us so
much. Everything plays its part and nothing is left to
waste.

This is the way of the natural world, and it is the

source of the wisdom that is at the heart of my people and many other indigenous peoples of the world. It is the reason that we have such a part to play on the international stage when it comes to respecting and saving our planet.

I have spent my life between the country and the city, and in Africa we find a value in everything. In the countryside, we do not throw spare food away. We give it to our neighbours, or we dry it to eat another day. In the city, we don't just dispose of cars, we turn the tyres into shoes, car doors into suitcases and wheels into cooking stands. Our philosophy is that everything has a purpose, though the recycling at this point is imperfect.

For waste to be fully designed out of the system, products must be designed to last longer, and at the end of their life they must be able to become a new product, as good as the old one. That's how you mimic nature – by making a perfect circle. Already there are the beginnings of some fantastic techno-logical solutions that can help make this happen. By combining the philosophy that every item is important with the technology of today, a waste-free world becomes more than a dream.

I believe we are in charge of our destiny and can change our relationship with our world. There is no reason why we can't combine the knowledge and experience of indigenous peoples with the talent and resources of the world's biggest nations to achieve what nature manages to achieve every day. If we bury the planet in rubbish, we will all be responsible, but I am determined that this is not going to happen. Around the world, we are finding ways to give waste a value and to turn it into something useful. People always want more stuff, but if we can produce locally, think about everything we buy, use our brains to re-design our products and our lives, we truly can create a world without waste. The potential for change is huge.

Just imagine if we could grow more food locally and eat everything before it rotted. If we didn't buy things we didn't need. If plastics and factory waste were not discarded into our rivers. If we could make every product last as long as possible, and waste could be designed out of the system. Nothing we produced would make it to the sea. No food would rot, and no one would go hungry. Dumping sites would become beautiful hillsides. Rivers would spring to life with nature. We could learn from Nature, and use no more than what we need, and reuse what we do need in a perfect circle. If we can do this, the world will be much more beautiful, balanced and plentiful, for both wildlife and humans, and we won't leave our foot-prints behind.

Hindou Oumarou Ibrahim, 2021

The urban streets of India, at first glance, may appear a mess – but they are an inspirational mess too, concluded Canadian historian Robin Jeffrey and Australian anthropologist Assa Doron, after disentangling the story of the country's waste. The mess, they say, is a mess that holds lessons for many 'cleaner' societies about the principles – if not always the practice – of minimising waste and recycling anything and everything.

Like most places, almost everything discarded in rural India was once biodegradable. Scavenging animals soon ate it, or the heat and torrential rains broke it down until it was all but invisible. Much of it soon made its way back to nature, fertilising soils and bringing new plant life. Now, courtesy of modern technology and our rampant consumerism, human-made non-biodegradable garbage is everywhere: in backyards, on street corners, floating in rivers and piled up across wasteland.

India is not alone in its failure to handle its growing piles of non-biodegradable garbage, of course. But it is more crowded than most. The country has almost as many people as China, living in a third of the area. 'Never in history have so many people had so much to throw away, and so little space to throw it,' say Jeffrey and Doron in their book *Waste of a Nation*.

But there is another side to this story. The piles of trash grow partly because nothing is truly thrown away. Almost everything in India has an afterlife. In time, someone is always there to pick it up and make a rupee out of it. The economies of many slums are built on handling the trash from the cities around them. They include Dharavi, the poor suburb of Mumbai made famous when it served as a backdrop for the 2008 movie *Slumdog Millionaire*. One of Dharavi's most thriving quarters is devoted to lines of workshops where workers sort and recycle the megacity's waste. In their hands, it is not waste at all. They chop plastic bottles into pellets, smelt aluminium cans into small ingots, restaple old cardboard boxes, wash cooking oil cans, mould used hotel soap into new bars, and sift mounds of garbage to extract ballpoint pens, metal jar tops, toothbrushes and much else.

Repairing things is a way of life in India. They have a word for it: jugaad, meaning frugal innovation. Almost every street has someone at the roadside repairing bicycles and sewing machines, or taking apart abandoned cars and trucks. And what cannot be repaired is recycled by the millions of poor Indians who devote their lives to collecting, sorting, processing and selling almost anything that is discarded by their richer fellows. India recycles more plastic than anywhere else in the world. More than 60 per cent of it finds a new use. Plastic is the stock-in-trade of the humblest street picker. Most of the plastic bottles and bags, pens and wrappers that they collect are chopped and heated and remoulded by countless backstreet businesses, to make everything from window frames to chairs.

Some of the supply chains are staggering. Human hair is extracted from gutters by children, supplied by pavement barbers or bought from temples where it is ritually cut by pilgrims. It ends up stuffed into mattresses or sent to China to be woven into wigs for sale in the United States. Jeffrey and Doron found one Delhi trader who exports 60 tonnes of hair every month.

India imports inorganic waste too. Many of the world's dead ships end up there. The port of Alang in Gujarat has a ship-breaking zone extending 18 kilometres along the shore. Not a lifebelt, not a bathroom fixture, not a rivet goes unclaimed and unsold. This is not just a graveyard for ships, it is a reincarnation of their scrap metal contents for new industries.

The task of handling waste in India can often be dangerous. We saw earlier how whole suburbs of Delhi are devoted to polluting and potentially lethal ways of recycling old computers and other electronic waste, and read of the perils of recycling lead batteries. Boatmen on the Ganges fish thousands of half-cremated human bodies from its waters every year. Chennai's hospitals sell their medical waste to self-employed dealers recycling drugs and equipment, regardless of the risks. Dump pickers, exposed to smoke and disease, have an average lifespan of thirty-nine years. But if the practice can sometimes be desperate, the principle may be sound. Almost nothing goes to waste. India may have a lot of trash, but it recycles like nowhere else. The world needs to learn the lessons.

Or rather, to relearn them. London and other big European cities used to recycle as India does today. Before municipal collections and waste transfer stations, London's foremost nineteenth-century tradespeople included bone pickers and rag gatherers, sweeps and cigarette pickers, with dredgermen and mudlarks on the River Thames – all collecting and recycling the city's trash. As soon as there were sewers, scavengers, known as toshers, were lifting manhole covers and sliding underground to see what had been flushed away. Nobody does that any more. But Londoners still flush a lot down there, including unwanted drugs and wet wipes that do not degrade in the environment, evade the best endeavours of sewage treatment works, and end up poisoning marine ecosystems.

So we need a radically different approach to our handling of waste. We need a new mindset to replace that of our current throwaway culture, in which we extract materials from nature or the earth, make products, use them once and then discard them. We need to reduce what we use, especially of what won't degrade. We need to resume the old tasks of refilling, reusing, repurposing and recycling what we continue to use. Some of that we can all do. Some of it will require making products from materials that biodegrade, or developing new technologies that degrade what is currently non-biodegradable, such as many plastics.

Nature sometimes has methods that we can exploit. It has been working on recycling for millions of years. The goal should be to mimic nature by making the waste from one process into the feedstock or fuel for the next. To have a waste-free world we need what economists are calling a 'circular economy', in which culling new material from the ground or nature is the last resort, not the first. The production line must become a production circle.

The veteran American environmentalist Lester Brown says in *World on the Edge* that, from a historical perspective, the throwaway society is very new, 'an aberration that has evolved over the last half-century', and which may now itself be 'headed for the junk heap'. Certainly, as American architect William McDonough puts it, waste and pollution are 'symbols of design failure'. An economy constructed around single-use products is not only wasteful of resources, it is also inefficient and expensive.

Some industries recognise this. Steel from recycled scrap takes only a quarter as much energy as new steel made from iron ore. That is why scrapyards are full of crushed cars awaiting a buyer, and why most steel in everything from Coke cans to bridge girders is recycled. Recycling aluminium requires just 4 per cent as much energy as making new, which explains why most aluminium ever refined from bauxite is still in use somewhere. Much the same equation holds for other materials that are recycled much less. Plastics that can be recycled typically need a fifth as much energy as making afresh. So why doesn't it happen?

• • •

Most of us like to recycle where we can. It makes us feel good, though we sometimes need nudging. Some countries recycle three-quarters of their household rubbish. Britain is stuck at 45 per cent. That is up from 20 per cent at the start of the century, but short of the target of 50 per cent once set by government for 2020. The US figure is lower, at around a third, but it does have success stories. Two-thirds of refuse in San Francisco gets recycled. But our trust in the virtues of recycling can be misplaced. In the real world, there can be alarming gaps between eco-efficiency, economic utility and social justice.

One issue is where the stuff we carefully put into our recycling bins ends up. In London, for instance, much of the glass doesn't make new glass. It is instead ground up and turned into construction materials. Many ardent recyclers are less than happy to know that their glass probably turns up in the foundations of new roads. But at least the construction is likely to be local. Much of the recycling produced in rich countries joins a vast global trash trade of sometimes dubious legality or ecological usefulness.

Scrap metal is Britain's biggest export to Bangladesh, where it often joins the cut-up remains of ships on the country's shores. Until recently, most of Europe's paper waste went to China to turn into packaging for the consumer goods sent back to Europe. China banned the trade in 2018, and India, Indonesia, Turkey and Vietnam are taking up the slack. Two-thirds of American paper and packaging waste is now recycled; but almost half is also exported, with Mexico a major destination. Plastic has a similar pattern.

Maybe some of that international trade makes sense. But stories of waste for 'recycling' simply being dumped in the countryside or left on docksides are frequent. And what qualifies as recycling is sometimes downright lethal.

Nobody should want their computer or mobile phone recycled as it is done in Delhi, nor sent to the Agbogbloshie dump in Accra, the capita of Ghana. Agbogbloshie has in the past decade become an international garbage dump. As many as 10,000 people regularly await the early morning arrival of trucks bearing the word's trash, before wading through the piles as they are disgorged. They are looking for marketable materials, including old TVs and computers, and insulated USB and other copper cables, which they burn to capture the copper for recycling, releasing clouds of toxic fumes in the process.

Doctors studying these dump scavengers find high rates of lung disease, chest pains, and toxins in their bloodstream and urine that have been derived from the waste they handle each day. This is a circular economy, you might say, but an unacceptable one. And not necessary.

When not being diverted to the international scrap trade, seeking the least-cost, most invisible 'recycling' option, much of Europe's e waste goes to high-tech metals recovery plants. One is the Boliden refinery at Rönnskär on the edge of the Arctic Circle in northern Sweden. Its copper smelter is one of the world's largest recyclers of copper and precious metals from end-of-life electronics. It also extracts tens of thousands of tonnes of zinc to produce zinc clinker and lead batteries. This is how it should be done.

• • •

Exporting waste for 'recycling' to poor countries via opaque international trading systems is asking for trouble, and

243

unlikely to produce the environmental gains we expect from our recycling. Plastics are becoming a big deal here. Plastic trash that was until recently seen as little more than unsightly litter is now recognised as a major ecological problem for the planet. Much of it doesn't degrade but lives on forever, broken into ever smaller fragments but never disappearing. Littered across the land, much of it eventually enters rivers and flows into the ocean.

Britain produces more than 2.3 million tonnes of plastic waste a year. It is the world's second largest per capita producer of plastic waste, after the United States. Half of that waste goes for recycling, according to the waste charity Waste and Resources Action Programme (WRAP). But more than half of that recycling is done abroad. Scandals have been endemic. Much is probably never recycled, but just dumped or burned. Policing waste enterprises is hard. Embarrassed by the stories of criminal mafias running the trade, governments of importing countries are pulling out of the business. First it was China. Then Turkey became a major destination, until it too announced a ban on imports of several key types of plastic.

The vast trade in plastic waste arises in large part because we use so much plastic, and nobody wants to pick up the pieces. We should curb demand. Putting a price on plastic bags in supermarkets helped a lot. But there are many ways. They include ending plastic packaging on fruit and vegetables in supermarkets, replacing single-use plastic bottles with refillable containers, and switching to biodegradable and compostable plastics. Greenpeace says European countries could halve single-use plastic by 2025, and by then also ban the export of all plastic waste, to both encourage domestic recycling and as a means of raising standards and driving innovation in ways of handling what we continue to use and discard.

Step forward John McGeehan. The Scottish molecular biologist is also a surfer who got angry at having to plough his board through plastic waste bobbing on the waves. So he starting looking for ways to get rid of it. He figured that nature might have an answer, even for these man-made substances. So he began looking in places where nature gets a long, cool look at plastic waste – inside landfills. In a Japanese waste site, he and local researchers found bacteria that produced chemical catalysts, known as enzymes, that ate away at the apparently indigestible. The processes were slow in the landfill. But when John got home he combined two of the enzymes and found that they worked much faster together, and did so at room temperatures. 'Learning from nature, and then bringing it into the lab', is what the academic at the University of Portsmouth calls it.

McGeehan's concoction works best on polyethylene terephthalate (PET), the plastic that makes up most plastic bottles and about a fifth of all our plastic waste. He is going into production with a big chemical company to mass produce the PET-eating enzyme. He is also developing plans for other formulations aimed at digesting other plastics. Other researchers have turned up bugs that eat polyurethane, which is rarely recycled at present. Perhaps one day they could create a cocktail of bacteria to eat their way through entire landfills. 'I believe we are just scratching the surface of what can be achieved with enzymes to tackle the wide variety of plastics that are polluting our environment,' he says.

Developing ways of digesting plastic products is smart, but it would be much better with methods of manufacturing plastics that make the task easier. And that means involving the plastics manufacturers. It turns out that just twenty companies manufacture half the world's single-use plastic polymers. There are some familiar names, including ExxonMobil, the world's largest oil company, Dow, the world's largest chemicals company, and Sinopec, the Chinese oil giant. Between them, the top twenty make 130 million tonnes of single-use plastic a year, of which only about a tenth is currently capable of being recycled. They have the problem in their hands.

Such a small group of companies might get together to come up with a better way. It could clear the oceans. The questions is, do manufacturers want to produce plastics that can be easily recycled? It could reduce sales, after all. Or can they be persuaded? Can we, in other words, create a global economy in which the rewards for doing the right thing are great enough – and the penalties for not doing so are high enough – to get the outcome we want and that the planet needs? This may be a job for lawyers as much as technologists. But probably we consumers will have the final say.

For we have to decide how we want to live our lives: surrounded by trash and single-use products, made in ways that rip up the Earth and cull nature; or in a 'circular economy', where we get more from less? Central to that will be how we run our cities, the places where most of us now live.

Cities have created many of the planet's problems. They are where our most challenging inorganic wastes, such as metals and plastics, are mostly produced and used. They occupy 2 per cent of the Earth's surface, but are home to over half the human population. Their inhabitants eat up three-quarters of the Earth's resources, and produce three-quarters of its pollution and waste. And yet, they have the chance to be incubators of new circular solutions. Cities are great places for collecting materials for recycling, for sharing services – think mass transit systems – for making buildings more efficient, and generally 'doing more with less'. They also generate new ideas – in bars, cafes and restaurants as much as in workshops and seminar rooms. And city authorities have the muscle to make these things happen. Through regulation and their own buying power, they can set new standards that others will follow.

Some are up for the task. Bill de Blasio, mayor of New York City, the twentieth-century icon of the modern world, home of the Empire State Building and Wall Street, reckons the Big Apple can catalyse the change.

These days, New York is not even in the world's top twenty largest megacities, but it claims to be the second largest in terms of consumption, with more than a trillion dollars of products and services sold each year. But undaunted, the mayor wants his fiefdom to be the world's most successful 'circular city' by 2030. His advisors tell him that by reducing waste and creating an economy that is 'restorative and regenerative', it can get even richer. Repair and reuse, repurpose and recycle, are his new mantras.

The city wants to do obvious things in the coming

decade, such as improving its trash recycling from a currently paltry 20 per cent to 90 per cent. It wants to create 'circular malls', where everything for sale is second hand: a giant jumble sale, you might say. It has start-ups trying these ideas out. The Queen of Raw, set up by Stephanie Benedetto, is an online marketplace for buyers and sellers of cast-off unused fabrics from the city's most chic fashion houses.

On the face of it, not much of this is exactly cutting edge. Once, almost every Briton drank milk from glass bottles that were emptied and left on the doorstep for the 'milkman' to collect and recycle. People returning soft drinks bottles to the store could claim a cash 'deposit'. Somewhere along the way, these simple methods of resource recovery largely disappeared. Recycling became what poor people did in Mumbai slums, or a niche activity for rich greens. Now it is being reinvented as a mainstream business. The transition towns we saw in an earlier chapter pioneered the new spirit. But let's not carp. If New York can reinvent its consumer culture to create a circular economy, surely anyone can.

But inventing circular economies in cities will require going far beyond a bit of domestic recycling. The city authorities, like most others, are missing the biggest mote in their green eyes: the city itself. While most citizens work hard to recycle glass bottles, cardboard packaging and even plastic bags, city authorities continue to ignore the potentially much bigger benefits to the planet of recycling the materials that make the city's houses and apartment blocks, stadiums and skyscrapers, highways and bridges.

Cities are constantly being remade. A single large building can produce more waste when it is taken down than a city's 10 million inhabitants produce in a year. The concrete and steel in urban construction make up most of the stuff we mine from the Earth, and their production consumes massive amounts of energy and emits getting on for a tenth of our carbon dioxide. But when a demolition team turns up on site, virtually none of the huge volume of rubble they create ends up being recycled.

Ideally, buildings should be designed so they can be taken apart without a wrecking ball, and their components used again intact. And there should be legal limits on the carbon emitted to produce each square metre of floor area in any new or renovated building. But even when not, 'many building materials can be reused, recycled and recovered', says Brian Norton of the Dublin Institute of Technology. Concrete rubble should not be dumped in landfills. It is not hard, it's just that nobody much is doing it.

This is a doubly curious blindness, when most countries are developing programmes for renovating buildings so they need less heating or air conditioning. Renovating a building to reduce its energy consumption makes little sense if there is no control of the carbon-intensive materials and components used. The energy 'embodied' in the construction of a typical building is the same as eight years of use.

• • •

If, thousands of years from now, a future species or alien visitor digs into the Earth's sediments and rocks in search

249

of clues about the people who lived there, they may discover a vast layer containing our trash, rich in inorganic waste and concrete rubble, electric goods and slowly rotting plastic. They would conclude that the Earth became for a time a giant landfill. Unless, that is, we decide to do something different. For a truly circular economy of the kind we surely need in the twenty-first century, we would not just need to find ways of recycling materials in current use, we would also need to work to bring back into use some of the billions of tonnes we have already dumped into landfills.

Right now, waste dumps are toxic time bombs. Forty years ago, after a scandal over high rates of cancer, leukaemia and miscarriage among people living on a filled-in chemicals dump called Love Canal in New York State, the United States government launched a national dump clean-up campaign. Called Superfund, it spent tens of billions of dollars, often removing dangerous waste and burying it elsewhere, hopefully more safely.

But one day soon, many dumps will be mined. Automated machinery will dig through them sorting materials for reuse and recycling. Spoil heaps left behind by old metals mines are already being opened up. Many contain materials in concentrations higher than the ore in currently worked mines. With modern mining technology, taking a second bite makes economic sense.

Many waste dumps around the world are probably at least as valuable. Waste authorities already remove material when space in the dump runs out, or to staunch the flow of pollution getting into underground water supplies. Now, some are starting to realise that they are sitting not on a troublesome trash dump but on a mine of valuables, whether car tyres or soil, bed frames or aluminium cans, iron nails or car parts.

The American dumps may have some of the richest pickings. More than 100 million tonnes of trash goes into more than 1,000 American landfills each year, of which around 1 per cent of the contents comprises metals that may be worth billions of dollars. The first serious dump-mining enterprise has been at a landfill in Maine. A scrap-metal company brought in magnets to attract ferrous metals and giant blowing machines to separate out non-ferrous metals. It successfully recovered more than 30,000 tonnes of steel, copper, aluminium and silver with a sale price of around 7 million dollars.

Travis Wagner of the University of Southern Maine, who studied the project, says in his research that the task was made easier because the landfill contained ash from an incinerator, which concentrated the metals. But even so, it was a model for 'the much bigger prize, and bigger challenge of mining raw waste'.

• • •

We need to get back to basics on waste. Concentrate on cutting our usage and recycling the things that really pollute the planet, rather than focusing on those that are easiest to recycle or which best match our lifestyles. That puts at the top of the list ending emissions of the carbon dioxide that is warming the planet. But right up there too should be fixing waste nitrogen. As we saw earlier, nitrogen pollution is one of the three dangerous planetary 'boundaries' that we have crossed. But while climate change and lost biodiversity are constantly in the media, we rarely hear about nitrogen.

Yet it is spilling from every field into every river and on into the oceans, wrecking ecosystems in soils and rivers, and ultimately creating vast ocean dead zones. It is pervasive and devastating, because it messes with the basic processes that create and sustain life.

Currently, the world wastes about 60 per cent of what is poured on to the land. It never gets near crops. Remarkably, this figure is higher than it has probably ever been. There are very few industries round the world where efficiency in the use of resources is getting worse – but fertilising fields is one of them. Nonetheless, so far, there is no equivalent of the UN convention on climate change to address the issue, and few environmental activists are calling out corporations on the matter. Farmers just keep pouring huge amounts of often heavily subsidised fertiliser on their fields in the expectation of bigger yields that are often illusory. Such days should be long gone.

The UN has set a target, known as the Colombo Declaration, of halving nitrogen 'waste' from chemical fertiliser by 2030. It would be a good start, though it would only take us back to the levels of waste of a few decades ago. Like most waste saving, it would probably reduce farming costs rather than increasing them. Yet you have probably never heard of it. And that is not surprising, because only thirty countries showed up to negotiate it, and many have yet to ratify it.

We could do better again if we reduced food waste. A third of what is so wastefully produced on the world's agricultural land is never eaten. It gathers in warehouses, wastes during transportation that can take it to the opposite side of the planet, or is thrown away by inefficient supermarkets or fussy eaters.

How can we stop more than 100 million tonnes of valuable and polluting chemicals going to waste each year, clogging the world's rivers and creating dead zones in the oceans?

We will discuss our food waste later, but first off, how can our farms work better? How can the UN target be achieved? Farmers have for years been lobbied by manufacturers and sometimes encouraged by government incentives into buying and spreading ever more fertiliser. They have to be helped to use it better, more economically and more wisely. In richer countries where precision farming is already starting to use drones and computers to predict exactly what inputs plants in fields need, and when, nitrogen is high on the list of inputs to be controlled. Drones can spot signs of blight that require treatment, and reassure farmers that at other times they do not need it.

For poor farmers, the immediate solution may be nitrogen in the form of packaged pellets that can be placed carefully close to roots. The nutrient will seep out of its casing gradually, directly dosing the plant roots. It is the nitrogen equivalent of drip irrigation of water.

For many farmers, organic methods will be best. These can include planting leguminous plants such as peas, clover and soy beans that contain natural nitrogen-fixing bacteria in their roots. If planted as part of a cropping cycle, they maintain soil fertility without chemical additives. That once widespread practice has fallen out of favour, because it means farmers have to leave their fields without their primary crops while the legumes do their work. But now crop breeders want to tweak the genes of the bacteria so they will attach to the roots of farmers' favourite crops, such as wheat, rice or maize. Trials in California show it works. As well as drastically reducing nitrogen run-off from fields, it would cut the carbon dioxide emissions from manufacturing the fertilisers.

There will still be some nitrogen pollution, much of it reaching the oceans. So in the spirit of recycling, maybe we had better find ways to soak it up before it kills ocean wildlife. Preferably profitably. The nitrogen is a nutrient, after all. Phoebe Racine of the University of California, Santa Barbara, reckons seaweed cultivation would work. She has her eyes on one of the world's largest dead zones, in the Gulf of Mexico, which is gorged with nitrogen pouring from fields the length of the Mississippi River. Growing seaweed in giant pens near the areas where dead zones occur could soak up this excess nitrogen, she says. The resulting seaweed could be fed to livestock, for instance. Perhaps, as we saw earlier, the seaweed could reduce methane emissions from cattle, or even be put back on to the land, further reducing the need for fertiliser.

That would be a smart way of helping fix the broken nitrogen cycle. Here is another smart way. Making good use of human excrement. For this too is a nitrogen fertiliser playing havoc with the world's water ecosystems.

Welcome to the world of the honeysucker. Don't be fooled by the name. Honeysuckers are waste trucks that cruise the streets of Bangalore, India's newest hi-tech megacity, sucking up its low-tech problem: sewage. The trucks empty Bangalore's million septic tanks and pit latrines, where the majority of its 10 million inhabitants relieve themselves. In most cities where sewage systems fail to reach households, such trucks take away the urine and faeces from tanks and discharge them, often in the dead of night, into streams and lakes. But in Bangalore, the honeysuckers can work in daylight, for they head to farms, where their stinking loads are in demand to fertilise vegetables and coconut and banana trees. Farmers pay good money for human waste; it produces bumper crops.

The honeysuckers are another example of India's ingenuity at finding a use for every waste. But they are also evidence that the world of excreta is being turned upside down. The realisation is growing that our faeces and urine are not simply waste to be disposed of as fast as possible, but a valuable source of nutrients that could be helping to feed the world.

Your digestive system is a small fertiliser factory with an output sufficient to nourish 200 kilograms of cereals a year. Properly applied, the world's excreta could replace almost 40 per cent of the nutrients in chemical fertilisers. So it is time to stop holding our noses and start celebrating the farmers who recycle sewage on their fields.

They are, after all, only reviving old ways. Traditionally, sewage has always been collected to grow vegetables and other crops. European cities in the nineteenth century were fed on sewage farms. Much of the sewage from Mexico City is still piped to the fields of the Mezquital Valley, where the slurry remains of the city's digested meals simultaneously fertilise and irrigate new crops, doubling yields and tripling the rentable value of farms. Poop has made the farmers of the Mezquital Valley wealthy. In Pakistan, sewage grows up to a quarter of the country's vegetables. In the Indian state of Gujarat, farmers compete for sewage at annual auctions. Now, the honeysuckers of Bangalore are offering farmers another option.

The biggest argument against agricultural recycling of sewage is that it carries disease. But maybe the answer is not in banning the practice but in making it safe, says Pay Drechsel of the International Water Management Institute in Sri Lanka. Storing the waste first in ponds, and sprinkling it with wood ash or rice husks, kills many dangerous pathogens. Better still, treat the sewage before giving it to farmers. A typical sewage works removes most pathogens while leaving behind most of the nutrients. Israel uses about 70 per cent of the treated effluent from its sewage works for irrigation. Much of the treated sewage from London is sold to English farmers as a cheap fertiliser. Even human waste can have a role in a circular economy.

PART 3

OUR EARTHSHOTS – HOW WE CAN ALL MAKE A DIFFERENCE

9

A Good Anthropocene

The first TV Tom Szaky ever watched was on a set he rescued from a pile of junk outside an apartment building on his street in Canada. The young Hungarian refugee from the break-up of Eastern Europe's Communist bloc in 1990 had never lived in a house with a television before. 'We're like, let's take it. We plugged it in and it worked,' he says. The joy he found in reclaiming that television set for another life has stayed with him. 'It lit this light bulb for me about waste.' He now runs a recycling company called TerraCycle that has dozens of big brands as clients, and offices in twenty-two countries – all of them furnished with upcycled waste, he says.

Leaving aside the television set, Tom seriously started in the recycling business when he was a student at Princeton University in New Jersey. He collected food waste from campus dining halls and fed it to worms, which excreted organic fertiliser. He packaged the worm poop in plastic bottles obtained from campus refuse collectors, and sold it to gardeners.

Studying business, he had already figured that waste was the industry to be in. 'Almost everything we possess ends up belonging to a garbage company,' he says. But those companies too often treat the waste as garbage that is expensive to dispose of, rather than as a potential raw material with its own value. If he could change that paradigm, moving from a linear to a circular trajectory for the world's waste, he reckoned he could transform almost any business. TerraCycle was born. Ever since, he has been thinking about waste, 'how to collect it, how to transform it into something new, and figuring out whose gonna pay the bill?'

Tom has developed processes for turning everything – from cigarette butts to soiled nappies and chewing gum to pandemic PPE – into new products. From the dirty nappies, he extracts the super-absorbent plastic that can help soils in dry lands hold on to moisture. From millions of discarded cigarette butts – which he collects in ten countries – the filters make new plastic products while the rest is turned into compost.

Tom's big business plan is what he calls 'the loop', a system that gives value to packaging waste. It hooks manufacturers and retailers into selling products in stylish reusable containers – coffee cups, shampoo bottles, burger packets and more – and then accepting back the packaging from all the other brands and retailers that are signed up to the system. Customers get attractive up-market containers. They pay a small deposit for the packaging, but get the money back when they return the container for reuse. Along the way, everyone – customer, manufacturer, brand owner and retailer – feels good about engaging in recycling.

'Loop will refund you the deposits. Then we take the dirties, sort them out, clean them and provide them back to the manufacturer, who refills them and sells them again,' he says. He is working with 150 major brands: Nestlé and Burger King, Tesco and McDonald's, Mars and Procter & Gamble among them. 'What it's doing fundamentally is shifting packaging from being your property as a consumer to the property of the manufacturer', who is then motivated 'to make the packaging more durable, and lower the cost of every use'. It's a great sell, and is making Tom rich. That doesn't embarrass him. It means others can get rich, too.

Tom champions reuse rather than recycling. 'Recycling is not an answer to waste,' he says. 'It's an answer to a symptom.' The real disease is society's habits of disposability. Fixing those, he says, means finding a way back to a world before the 1950s. To a world where 'we cobbled our shoes, we mended our clothes, and where everything from our perfume to motor oil came in reusable packaging'. A world in which a typical shopper 'bought two apparel items a year and used them for twenty years until they became rags', rather than one where we 'buy sixty-six apparel items a year and wear them an average of three times before disposal'.

He believes the way forward is to concentrate on changing the world rather than the people. 'Let's accept people for who they are. We could sit here and vilify people for being selfish, or we can say that's how it is.' Incentivising them to change their ways is 'the real challenge for the sustainability movement'. In that, 'every actor has a role', he says. 'Governments can pass legislation that makes it difficult to do linear things, by banning straws, plastic bags and disposable takeaway food packaging. Corporations can create circular options . . . and retailers can amplify the good choices and reduce access to the bad choices.'

We cannot all hope to make a difference on the scale of Tom. But he is one impressive role model for anyone with ambitions to change the world. And he is far from alone. This book is full of people with a personal vision that changed things on a much bigger scale. Here is another.

Back in 1977, John Grimshaw got so fed up with cycling on dangerous roads in the English city of Bristol that he formed a group of fellow cyclists to create safe cycling routes. They bagged an old abandoned railway route to nearby Bath, and were away. Sustrans was born. It kept finding old railway tracks and other byways, and persuading local authorities to convert them into cycleways.

Today, Sustrans is a national organisation with thousands of volunteers in charge of a national cycle network of 20,000 kilometres – or Land's End to John O'Groats twenty times. Meanwhile, cycling has soared nationally, as people get out of their cars and onto their saddles, swapping petrol for pedal power. In an earlier chapter we saw how just a generation ago Beijing was full of bicycles, and cars were a rare sight. Imagine something similar returning as the global norm.

The Dutch have always done it. Almost as many Dutch people cycle as drive each day, especially in cities. Holland's train stations are dominated by vast cycle parks rather than car parks. For a long time the flatness of the country made cycling more popular than it was elsewhere, but with electric assists growing in popularity, hills are no longer the problem they once were. This could be just the start of a cycling revolution. Started by people like Grimshaw.

· · ·

Sometimes a group of enthusiasts working together can be transformative. It can be especially powerful if those enthusiasts include both consumers and producers, joining forces to change whole industries. Take guitar making. Some may think this to be a niche business. But 2.6 million guitars are made every year. Guitar makers are particular about the wood they use, because tonal qualities and reverberation can depend critically on the quality of this material. Often that has required using rare wood with a dense grain, provided by old trees with a large girth – many of them endangered.

Sourcing such wood can be difficult. To meet those demands, US-based Martin Guitars, in Pennsylvania, claims to have thirty suppliers on six continents. Making sure the wood is cut sustainably and legally can be even harder. Classical and acoustic guitars have traditionally been made with Sitka spruce soundboards. The world has plenty of spruce, but far fewer of the 400-year-old trees that make the best instruments. Necks are often made of mahogany, and acoustic backs and sides of rosewood, for which supplies are running low and the trees are usually protected. Big-leaf mahogany, one of the most prized woods for guitar bodies, has been on the endangered species list for two decades.

Traditionally, guitar makers have been secretive, almost cultish, about how they source their timber. This has sometimes been a cover for illegality and ransacking of rare woods. There have been scandals. A decade ago the American company Gibson Guitars was raided over allegations about the legality of its supplies of Madagascan ebony and rosewood. The company said it wasn't doing anything wrong, or nothing that others weren't doing too. But in a settlement, it paid a large fine, forfeited a lot of wood and agreed to change its ways. In the wake of the scandal, guitar makers and their customers have been trying to clean up their industry.

Leading musicians including Mick Jagger and Sting, Willie Nelson and Lenny Kravitz, and bands Maroon 5 and Guster have been calling for sustainable sourcing. Online guitar seller Reverb teamed up with the Environmental Investigation Agency to root out the bad guys. Manufacturers are falling into line. That is giving a boost to sustainable producers of wood, such as the Forest Stewardship Council (FSC) and Rainforest Alliance-certified harvesters of mahogany in the community forests of the Maya Biosphere Reserve in Guatemala.

Guitar makers are taking proactive steps to secure sustainable supplies. Taylor Guitars pays for ebony replanting in Cameroon. The search is also on for alternative timbers to break the mystique of the old woods. Australian guitar makers are picking up on the acoustic qualities of timber from previously ignored native species, such as Tasmanian blackwood, Queensland maple and bunya pine. Americans are buying guitars made of acacia koa wood sourced from the slopes of the Maui volcano in Hawaii. Others are into salvage timber.

. . .

Ultimately, the pressure for change often comes from us. As customers, we demand it. As voters, we demand it. As shareholders, pension fund contributors and employees, too – we demand it. We don't want to buy products with a hidden sticker saying it comes 'with added planetary destruction'. We don't want our money or our working lives devoted to helping the bad guys. Many corporate executives these days say an important reason for having a green image is to attract the best graduate recruits.

So, how can we use this power? Of course the best purchase, as Tom puts it, is no purchase. But if we need something, how can we have it with the least environmental impact? Chester Zoo's orangutan conservator Cat Barton, who we met earlier, says: 'We can't ban palm oil, but if we all demand it, we can have deforestation-free palm oil. As consumers we can really help to create demand for more sustainability. People power makes a huge difference, and we can work together to save tropical forests just by looking more closely at what we're buying.'

She has successfully campaigned to make Chester a 'sustainable palm-oil city'. But however strong our motives, most of us cannot spend our days checking the back story of potential purchases. So we often need NGOs or other investigative agencies to do the heavy lifting on our behalf.

With this motive, a range of certification schemes have been set up, where producers and environmental groups thrash out some ground rules for supplying markets in ways that do not cost the Earth. Among them, as we have seen, are the Forest, Marine and Aquaculture Stewardship Councils, Fairtrade and Rainforest Alliance labels, the Soil Association and the Roundtable on Sustainable Palm Oil. Such labels involve compromises to avoid making the perfect the enemy of the good. Sometimes, technologies divide even otherwise like-minded people. Is genetic engineering of crops an acceptable way of increasing crop yields and 'saving' wild lands from being cultivated, or are there downsides that should keep it out of bounds? How important should social justice and land rights be in an environmental certification system? Are big food combines such as Unilever, co-founder of the Marine Stewardship Council and the Roundtable on Sustainable Palm Oil, part of the problem or part of the solution?

This is all understandable. Nobody said saving the world was going to be easy. But we can't afford to walk away. There is no going back. The Earth and the way its life-support systems function have been irrevocably transformed by our activities. Over the past couple of centuries we have moved vast amounts of carbon from the Earth's crust to the atmosphere, and similarly large quantities of nitrogen from the atmosphere into the land and oceans. We have to stop doing that, but there is little chance of going back to the old world.

There is no escaping the fact that we are in the Anthropocene: the new geological age in which humans dominate everything. We have no choice but to take charge of our planet's destiny, because it is our destiny. 'We broke the world; now we own it,' says Ted Nordhaus of the Breakthrough Institute in California, in a recent blog post. 'There is no alternative to actively managing the Frankensteined Earth systems that we have unwittingly created.'

Planet Earth is our home. Nobody trashes their own home, burning the furniture, blocking the drains and blowing the roof off, but for a while we have treated our planet in just that way. We had a big party. Now is the morning after. We need to sober up and clean up. As we saw in Part 2, many of the solutions have already been developed. Some are already happening. We need to celebrate them, and scale them up to make what now seems radical, brave and 'green' into the new normal.

We are optimists. The good news, we believe, is that we can end waste, fix our climate, replenish the oceans, restore nature and clean the air. Our good Anthropocene will be a world where our planet and its goodies are treasured rather than squandered; where carbon is kept in the ground because we have broken our addiction to fossil fuels; where we feed ourselves without clearing forests and ploughing up grasslands; where we mine waste dumps and not geology; where transport does not cost the Earth; where the Earth's natural systems are rebooted. We can have abundant clean energy and wild landscapes, thriving seas and well-stocked fishmongers, cities worth living in, air worth breathing, stable climate and shorelines, circular cycling of resources that banishes waste, and food grown from hydrogen and air.

This will often be a high-tech world, where technology brings ecological as well as economic efficiency, allowing better human lives and more room for nature. But it will also be a world where ancient wisdoms – respecting nature, husbanding resources, and sharing between us what is best shared – are rediscovered.

Many of the solutions we need will also chime with the equally vital goal of making life better for the world's disadvantaged people. Despite many advances in human well-being, our world has never been as unequal as it is today. Many believe the roots of that crisis are the same as the roots of the environmental crisis. Surely, they need common solutions.

So tackling the five Earthshot challenges will dovetail with many of the United Nations' 2030 Sustainable Development Goals. The key word is 'sustainability'. Abstract though it sounds, and much as it has been abused, 'sustainable' has a real meaning in making things better for people as well as for the planet. To take one example, improving the lives of billions of the world's poorest won't increase energy use. Mostly, it will reduce it. As Marta Baltruszewicz of the University of Leeds has shown in detailed research in Nepal, Vietnam and Zambia, the most deprived often use more energy than the less deprived, while improving the lives of the worst off reduces their energy footprint by 60 to 80 per cent.

How come? Because without electricity or gas, they burn more firewood, with resulting pollution and deforestation;

without clean drinking water, they constantly boil dirty water to make it safe; and they are unlikely to be near public transport systems or other means of sharing resources. For the very poor, lots of basic things cost more, waste more resources and cause more pollution. Reducing their deprivation results in lower energy use and reduced environmental impacts.

So it is not only just and fair for poor people to have better lives. It is also ecologically vital, because it allows them to better protect their surroundings. And as we have seen, ensuring that indigenous peoples have control of their traditional territories is the best way of protecting those territories. Interlinking these human and ecological strands of justice and well-being is, we believe, fundamental to making a good Anthropocene.

Time is short. We have to get a lot of things right in the next decade. But we can do it, if we keep our eyes on the prize, going forward and not back. As environmental journalist Elizabeth Kolbert has it in her book *Under a White Sky*, the choice we face is 'not between what was and what is, but between what is and what will be'.

• • •

Some say this vision of a good Anthropocene is naive. They say the world is doomed because there are too many of us and we consume too much. There are two replies. First, the population explosion is coming to an end; and second, we can produce what we need so much better than we currently do.

Human numbers are still rising. We are well on the way

269

to 8 billion, four times what the figure was a century ago. But the number of children an average woman has in her life, known as the fertility rate, has fallen to 2.3, less than half its peak in the mid twentieth century, and falling close to the long-term replacement level of around two children.

The fertility range between countries is at present huge. In Niger, women average close to seven children, and in ten African countries they average more than five. But in most of the world the average is below replacement level, and in Taiwan, South Korea and Singapore it is below 1.2. Of the world's two most populous nations, India is at 2.3 and China at 1.3. Almost everywhere the trend is strongly downwards. Africa is the last continent into the decline, but it is happening there too. Half a century ago, Kenyan women averaged eight children, now they average 3.3.

Nothing is certain. But most analysts say that if fertility rates continue to fall, we may reach 'peak population' later this century, at around 10 billion, with numbers stabilising or even falling after that. Furthermore, there is clear evidence that wherever women are given equal rights and control over their lives, wherever more investment is put into access to health care and enabling women and girls to stay in education longer, birth rates fall. We should invest in these things across the world regardless, because it is morally the right thing to do, but if it also ultimately leads to a more sustainable world, too, then that is a double win.

But while it is possible to see the prospect of a stable, albeit much larger population, we also have vastly unsustainable levels of consumption. Even if some nations have, by some measures, reached 'peak stuff', there is a lot of understandable ambition for catch-up elsewhere. Still, as we have seen in Part 2, there is vast untapped potential for the world to produce everything from energy to food, and electronics to transportation, in ways dramatically better than we do today. Though the mountain looks huge, each one of us can champion solutions or alter habits in ways that accelerate the change that the world needs. If we do things right as individuals, then we can help the whole system to shift and ultimately the Earth can supply the needs of all of us. That is the prize in the Anthropocene.

· · ·

Can we afford to do all this? Some say sustainability is too expensive. But that misses the whole point of being sustainable. It means something we can do forever. Our current unsustainable practices simply mean we don't pay the bill until later. So the straight answer is that we cannot afford not to. The downsides of business-as-usual are ever more severe, while the ecological benefits of a new approach are ever more beneficial. One problem stifling progress can be our inability to measure all this. Ecological costs and benefits can be hard to compare with economic ones. But you can, as the business saying goes, only manage what you can measure. So environmental economists have spent years devising ways of putting a dollar sign on the 'ecological services' that nature provides – whether it is storing carbon or protecting river flows, pollination crops or recycling nitrogen, controlling pests or preventing the spread of animal diseases to humans. There is no fixed number, of course, but they reckon that on average every dollar spent on ecosystem restoration will bring an eventual return worth ten dollars.

That is fine in theory. However, in the real world of business and finance decision-making, the issues are usually framed not in terms of long-term collective gain, but with an eye to who wins and who loses in the near term. If no individual, or corporation or national exchequer can profit from some investment in ecosystem restoration, then even a return for the wider world of 1,000 per cent may not be enough to ensure the investment is made. Unless we can organise things through laws or economic incentives, so that someone somewhere has an incentive to do the right thing. From Tom's 'loop' to carbon taxes and the shaming of corporations by environmentalists, a lot of energy is being expended in finding the necessary carrots and sticks to make what we all know is needed happen.

We see signs that the right things are starting to happen. Or at least, the right noises are being made. Business norms are being transformed. From factory floors to farm fields, we see examples of sustainability becoming the new mantra. Evidence is growing that eco-efficiency can translate into economic efficiency. Investment in storing carbon in soils can improve farm yields. Protecting coral reefs can nurture fish stocks as well as carbon and other biodiversity. Rethinking waste as a raw material can turn economic burdens into profits. Restored forests can protect local crop yields as well as the global climate. Cleaning up city air can attract new businesses and improve productivity, as well as protecting lungs.

Smart entrepreneurs and financiers, insurance companies and commodity suppliers are making the links. They say they want to maintain and restore the 'natural capital' on which future profitability as well as our planet's future depends. In 2021 more than 400 asset managers, bankers and investors controlling an estimated one-third of the world's bankable assets – from oil wells and coal mines to factories and plantations and airlines – signed a joint statement calling on governments to stop supporting fossil fuels. 'We stand at the beginning of a pivotal decade in which institutional investors and government leaders worldwide have the power to raise ambition and accelerate action to tackle the climate crisis,' they wrote. They know that not doing so will be bad for their bottom lines as well as the planet and the rest of humanity.

There remains a big gap between rhetoric and reality. Environmental NGOs found that many of the signatories of the above statement were major investors in climate-destroying industries. The British signatories alone were responsible for almost double the UK's own annual carbon emissions. New fossil fuel projects were still going ahead, among them gas extraction in northern Mozambique, oil wells in Suriname, oil and gas deposits in the shales of Texas, coal in southern Bangladesh and oil in Norway's Barents Sea. The UK government has pledged to halve the country's carbon emissions, but it still gives out more subsidies for coal, oil and natural gas industries than it does for renewable alternatives

Turning the juggernaut that is the global finance system is proving hard. Still, some things are changing. Peak coal happened in 2013. The fuel that drove the industrial revolution of the nineteenth century is now in what most analysts believe to be terminal decline. And we may be at peak oil, too. The fuel that more than any other made the world we know today has started to lose its hold on our lives, our economies and our politics.

There may be a lesson here. It happened almost without anyone noticing, with no grand declarations, government pledges or international treaties. In 2020, as pandemic lockdowns emptied our streets and our skies, there was a 9 per cent drop in the amount of oil the world burned. Many people assumed it would recover rapidly. But two other things happened. First, the rise of electric vehicles (EVs) that do not burn oil. And second, the prospect that, thanks to video conferencing, many business seats in aircraft might never refill. If either or both these things happen, then we will indeed have reached peak oil. And the decline could be fast.

In a recent blog post, Mark Lewis, head of sustainability research at asset manager BNP Paribas, has cited 'three Ds' driving down oil demand. The first is the decarbonisation of economies to meet the demands of the 2015 Paris Agreement on climate change. The second is deflation of demand for oil as electric vehicles and renewable electricity generation kick in. The third is detoxification, as cities horrified by growing evidence of the death toll from dirty city air, and emboldened by the experience of clean air during lockdown, curb exhaust pipe emissions.

Automobiles currently consume almost half the world's oil. So the elimination of petrol-driven vehicles could halve oil demand. Governments support the growth of electric vehicles primarily as a pain-free way to cut carbon dioxide emissions. How well that works will depend almost entirely on how the electricity they require is generated. If it is made from burning coal or natural gas, then the climate benefits may be minimal. But the rise of renewables offers hope that they can help drive down emissions as well as delivering peak oil.

British energy giant BP's annual energy review, widely regarded as the industry bible, has for several years been predicting that the world may reach peak oil in the 2030s. But in its 2020 review, it dramatically moved the goalposts. Of three future scenarios it analysed, global demand in two of them peaked in 2019, with only a tiny increase for a couple more years in the third. The company's analysts forecast a decline by 2050 of anything from 7 per cent under 'business as usual' to 70 per cent if – as company executives say they hope – the world heads for net zero carbon emissions.

By that time, most analysts say that aviation fuel will be much the biggest demand for oil. So if the world can cut out aviation emissions – with, say, green hydrogen – we may be getting near to what the world needs. Zero emissions.

Investors will be key. They are already pulling their money out of coal mines and power stations, fearing that they are rapidly becoming 'stranded assets' with no chance of making money. The same could soon happen to oil wells and pipelines. If so, then the money will disappear and things will change fast. Especially if their money goes instead into redoubled efforts in low-carbon technologies.

Having created a monster, the forces of capital may soon slay it.

• • •

We cannot necessarily rely on sense to prevail, however. There are bad guys out there, with hired hands carrying guns as well as dollars. Listen to one story of what can happen when they are in charge, when the high-minded words of environmental economics or ambitions for a good Anthropocene matter not a jot. This is the story of Berta Cáceres, for whom the price of defending nature was the loss of her own life.

They came for her late one evening, as she prepared for bed. It was March 2016, in the market town of La Esperanza, in Honduras. A heavy boot broke the back door of the safe house Berta had recently moved into, after threats. Her colleague and friend, Gustavo Castro, heard her shout, 'Who's there?' Then came a series of shots. He survived. But the most famous and fearless environmental activist in Honduras died instantly. She was forty-four years old. It was a cold-blooded political assassination.

Berta knew she was on a hit list. Everybody knew. She was the national face of a campaign against the proposed construction of the Agua Zarca hydroelectric dam on the River Gualcarque, a wild mountain river sacred to the Lenca people. And the dam's supporters would do anything to silence her. Berta had told her colleagues to prepare for life without her. 'I knew she was afraid,' fellow campaigner María Santos Domínguez, who lives in the remote indigenous village of Rio Blanco in the country's mountainous west, said a few months later in an interview with one of the authors for *Yale Environment360*. 'It was too much for her. I could tell.'

María has herself been attacked with a machete as she walked to her children's school. But, despite the deep scar on her face and Berta's murder, she was undeterred. The Lenca people, she said, 'were born here. It is our land and our river. If we lost the river, we'd die. We need its water to bathe, for fish, for water, for our crops and animals.' She often bathes her children in the clear, cool mountain waters where the dam was planned. 'The river is sacred to us. We believe in the spirits in the river and they give us strength to fight the dam builders,' she said.

Berta had been born into one of the most prominent Lenca families. She had helped revive their identity, based on defending their mountains and rivers, forests and plant life. She had established a training centre, nicknamed 'Utopia', and a network of radio stations. A year after her slaying, La Esperanza was still full of graffiti declaring '*Berta Vive*', and shrines were visible on street corners.

Eight people with alleged links to both the government's security services and a Honduran company, Desarrollos Energeticos SA, behind the dam project have since been convicted of being involved in Berta's murder, including in July 2021 Roberto David Castillo, the former head of the dam company, who was found guilty of coordinating the operation. Her death is one of many hundreds of killings of environmental and social activists in recent times; of people who have taken a stand for nature and local people, against the construction of dams and against mines, logging and agricultural projects in Honduras, a country the international human rights group Global Witness once called 'the deadliest country in the world to defend the natural world'.

Will the Agua Zarca dam ever be built? Some now doubt it. Following the outcry over Berta's death, European funders pulled out of the project. Five years later, engineers who had prepared the site for construction had not dared return. The vigilance of her people continues, and the battle for the future of the Lenca, their lands and rivers, goes on. Maybe even here, the corner is being turned. But it has been at a terrible price.

• • •

If the tectonic plates are shifting, the real issue is whether they are shifting fast enough. If we are to achieve what Prince William calls 'a world with clean air, healthy oceans, a stable climate, where nature can thrive and nothing is wasted', then we have no time to spare. As we have seen, many of nature's tipping points – on climate and much else – may be crossed in the next decade, if we continue with business as usual. So, we need to trigger good tipping points in human society before nature unleashes its bad tipping points.

We have seen many examples of successful transformation from bad to good, including the inspirational power of individuals to make a difference locally and globally. Juan Castro's campaign for a no-fish zone around the coral reef at Cabo Pulmo, in Mexico, became a template for the growing movement to create marine protected areas worldwide. Tom's worm farm at Princeton turned into a 'loop' to ensnare New York City's garbage and capture the world's biggest brands. The eco leanings of a Costa Rican president have become a beacon for global reforestation.

These stories sustain our optimism. Optimism that humans can change; that technology can allow us to do more with less; that, if given half a chance, nature is programmed to fight back and restore itself. But we are living on borrowed time. We may have only a decade to fix many things.

That means we need lots of people with lots of ideas. So let's all stop being pessimistic or denigrating humans as planetary pariahs. Let's become optimists, celebrating and reinforcing the powerful ability of humans to do good. Let's embrace the idea that individual effort can unleash exponential change, to make a better world.

Still not convinced? Let's try an analogy. Think of your own idea for change as a single lily pad in a vast lake. It looks like nothing. It is barely visible, occupying maybe a square metre in a lake of 500 square kilometres, more than thirty times the size of Lake Windermere. But your lily can reproduce – your idea, if shared, can spread. Perhaps it will double to two lily pads in a day, and then double again to four the next day, and keep on doubling.

So what happens? By the tenth day of doubling there will be 512 lily pads. They still occupy only a tiny corner of water, just a millionth of the lake. But doubling continues. Even by day twenty-nine the lake will only be half covered in lily pads. But by then the speed of change will be unstoppable, and the next day a final doubling will fill the other half of the lake. Suddenly, it is covered in lily pads. In a month a simple daily doubling has taken the number of lily pads from one to half a billion, enough lily pads to cover the entire surface of that giant lake. On the last day, a quarter of a billion lily pads were added. That's exponential transformation. That's the kind of change we need. And every day matters.

10

Your Ten-Year Challenge

The take-home message of this book, and of The Earthshot Prize, is that optimistic, determined people make a difference. That ideas, actions and inspiration can be transformative. We cannot all be technical innovators. But we can all play our part in changing the world: through how we live our lives, how we communicate, and perhaps most importantly how we join and collaborate with others to shape our shared communities. Repairing our world will take more than fifty prize winners over a decade; it needs millions of Earthshots from millions of people.

It can feel overwhelming trying to change everything at once, but if we all asked ourselves what we could do year on year over the next decade, then suddenly it becomes more achievable. For example, most of us need to halve our carbon emissions this decade to play our part in tackling climate change. That sounds impossible, but really it works out at about 7 per cent each year.

Most of us can imagine cutting our emissions by 7 per cent this year. We would probably be confident about cutting them by a further 7 per cent the year after, and the one after that.

Some changes are easy, and we can do them now. Others need a community or workplace or school to help change. All will gain momentum if our country's electricity is made cleaner, thanks to advances in low-carbon renewable energy; if public transport improves, thanks to increased demand and campaigning; and if governments act on promises to drive emissions down to net zero.

The aim of this final section is to give some suggested answers to the often asked twin questions of 'What can I do?' and 'Will it make a difference?' In particular, when answering the second of those questions to convey that while all of our actions will make a difference, it is often more about how our personal actions change the prevailing attitudes in society, thus unlocking the potential for rapid systemic shift, than it is about feeling guilty over our footprints.

We will give suggestions of immediate things you could do but also how ultimately as societies we need to change the bigger systems of business and government. Our hope is to help catalyse a move from the despair or guilt associated with environmental actions to optimism and momentum. The important thing is to start. What may begin with a few lifestyle changes might spark the thought that changes your business or community.

If we want to change, then the first step is to be honest about our own place in the world, not as an exercise in guilt but as a reality check and a starting gun for action. According to a UN study, the world's wealthiest 1 per cent of people together produce *double* the entire carbon footprint of the poorest 50 per cent. Think about it. That is a staggering disparity, and probably unprecedented in human history. But you don't need a luxury yacht, private plane or mansion to join that elite. Given the level of world poverty, many of us who live what has become a relatively normal Western lifestyle are probably much closer to that top 1 per cent than we would think.

Still, we shouldn't succumb to guilt, or condemn others too easily. We should recognise we all start from somewhere. We all live different lives, and have different ambitions and different possibilities. We should also recognise that many people around the world have the need and right to increase their consumption but also that rich people can be part of the solution too, as green innovators or investors, inspirers or legislators. What matters is whether we choose to be part of the problem or the solution.

So, if we want to be among the latter, what should our personal Earthshots be? How can we as individuals address the five great crises? There are no single answers. There is no blueprint. We will all make different choices, and that is as it should be. That is being human. But what follows should help inspire you, and help you make informed choices.

Some people enjoy numbers. If you are one of them, decide some things that matter to you, and get counting. The next few pages will help you get started. Most of us will want to reduce our impact on the planet, and will be seeking clues about how to do it. Check out the book's two countdowns: 'Your Carbon Footprint in Numbers' and

'Your Water Footprint in Numbers' (pages 327 and 329). These will give you an idea of what matters most. There are also websites by WWF, the Global Footprint Network and others to help each of us calculate in more detail our personal carbon footprint (think electricity, home heating, cars and planes), as well as others such as our water footprint (think cotton clothes, rice and meat), our nitrogen footprint (think almost all food crops) and our plastics footprint (think packaging waste).

Besides checking our current footprints, the calculators will do the maths to help us decide what we need to change to hit our targets. Choose your own targets, but if you are in a wealthy, high emitting country, a first step could be to target 7 per cent less each year until you have halved your emissions in a decade.

Some people hate numbers. That's fine. Follow what angers or inspires you. If waste annoys you, start by just buying less stuff or resolving to buy things that last, repairing stuff, buying second hand and giving away what you don't need through online groups or charity shops. For what's left, recycling should come into play. In many places now, recycling schemes handle most things, but if not in your area then you may need to initiate some community action. Join others at school, work or any other community you are a part of. The bottom line is, throw as little away as possible.

• • •

It is easy to believe that our environmental imprint begins and ends with carbon. But as we have seen, the planet faces other crises. Nature is in freefall, with lost species and diminishing ecosystems. Rivers are running dry in many parts of the world. And nitrogen, plastic and other wastes are flowing from our fields and sewer pipes in such quantities that freshwater ecosystems are dying and dead zones are forming in the oceans. Each is a planetary 'boundary' that we are crossing, with potentially grave consequences.

It is easy to see humans as a problem, even to see ourselves as a burden on the planet. Please avoid that. Of course, we should all cut our damaging footprints, but also embrace our potential and that of others to do good. We can have a good footprint, too.

The trick is to start right now, but take the long view. If the thing that excites you is restoring nature, then start by planting some flowers or letting the lawn run wild. Then ask what you would like to see ten years from now in your street, your neighbourhood and your town. And think about how you could start that happening. So, as well as rewilding your garden, you could campaign to restore nature in some other corner of your world. Maybe by planting native trees or nurturing a patch of wild meadow in an urban park or on an abandoned wasteland.

Reinforce your good footprint by spreading the word, by setting a good example, by inspiring people through your enthusiasm, and by coming up with new ways to do things.

The good Anthropocene will be achieved more by building things up than tearing things down. By creating a better world rather than just dismantling the old one. We cannot banish coal and oil from the commanding heights of our economies without having low-carbon, non-polluting renewables to put in their place. We cannot protect nature and the oceans without innovating in agriculture so that we need less land and create less pollution. We cannot banish waste without rethinking the entire resource cycle.

We need to keep our wits about us. As environmental enthusiasm grows, ever more companies are making misleading claims about how 'green' their products or services are. Often they do this by mentioning one aspect where they show up well, while ignoring others that may be more important. Instant clothes made of 'natural' fibres don't look so environmentally friendly through the prism of their hefty water footprint. An SUV isn't 'green' just because it has a hybrid engine. So always think for yourself, and check if those making such claims offer some independent support for them, such as carrying one of the labels we have mentioned earlier.

Things can appear complex and daunting. But the key thing is not to drown in guilt. We should do our own research, but at the end of the day we should do something to feel good about. Then share our experiences. Not to boast, but to share ideas, tips and above all enthusiasm. The stories in this book offer inspiration for individuals to achieve that good footprint – but also, we hope, to do the same for communities, peer groups, governments and whole societies. Politics matters, because politicians pass laws and make decisions in response to pressure put on them by voters, consumers and businesses. Make sure your voice is in there.

One person will rarely change the world alone. But one person can light the spark that changes a lot, or can carry a torch lit by others. So be an inspirer and an enthusiastic follower. It's of course true that many of our biggest problems need governments and the business community to make the biggest changes. Saving the world cannot be achieved by individual actions alone. But each of our actions – and our collective attitudes – ultimately determines what is normal or acceptable in a society. We can make the political weather that will ultimately decide the real weather.

• • •

The Earthshot is a ten-year project. Scientists tell us we have just a decade to make the fundamental transformations in our lives and our societies. There are things many of us can do ourselves right now. There are things we can personally work towards over the decade. And there are things that require collective action – at street or community, metropolis or corporate, government or intergovernmental level.

In each of the Earthshot themes that follow, we try to disentangle what each of us can do in the short and long terms, both individually and collectively. A ten-year target could be a personal one, such as 'A decade from now, I choose to end my own contribution to air pollution completely.' Or a community one, such as 'A decade from now my town will have more cyclists than cars on the road.' We cannot tell you how to achieve those things, but we can give you information and encouragement while you are on the road. Think of this book as a carrier of water bottles for a marathon runner. It can't run the race for you, but it can help.

This section offers ideas for potential personal contributions for each of the five Earthshots. Life and nature are not so simple, of course. There is a lot of crossover, but the good news is that most of the good things we can do will help in multiple areas. A more plant-based diet will help nature by saving wild lands and reducing water grabs from rivers, while at the same time reducing your carbon footprint and cutting nitrogen pollution of rivers and oceans and being good for your health. Driving less and switching to an electric vehicle will clean up air pollution as well as help reduce climate change. As will rewilding your garden or street. Consuming less of almost anything protects nature and the climate, while also reducing waste and keeping plastic out of the oceans.

Nature knows no boundaries. The planet works as a single system. Treat it with care, and imagine the world we can create with a decade of action.

Air Pollution

Briefly, at the height of the Covid-19 pandemic lockdown in 2020, many of us experienced what cities could be like when they are free from the tyranny of traffic and the fumes it produces. The air was clean, silence reigned, you could cross a road at leisure and hear the birds sing, wildlife returned and parks blossomed. The challenge is to return to that environment while not sacrificing productive economic and social lives. To create new, greener cities.

The dawn of electric vehicles offers that chance. By 2030 almost all new vehicles should be electric, with no kerbside emissions. The next challenge will be to reduce traffic and get rid of the surviving old polluting vehicles and their emissions. Beyond that, we can imagine urban landscapes of neighbourhoods fit for walking and cycling, filled with buildings that require much less heating and cooling, and which may even sprout their own vegetation. We can imagine remaining motorised transport increasingly confined to autonomous electric vehicles travelling on smart highways. The streets will be ours again.

A world free of polluting emissions will also be one in which other forms of air pollution are banished. They range from local threats – such as the particles that may continue to fly off vehicle tyres, and industrial emissions that cause localised smog – to the often invisible and odourless gases that spread on winds from the tropics to the Arctic. No more long-lasting toxic pesticides from farm fields, no more ozone-eating industrial chemicals. When they are all gone, we will truly be able to say the air is clean. We won't do all of that in the next decade. But we can go a long way. So how can each of us help?

FIRST UP

Most air pollution, especially in cities, comes either from the individual choices we make in our lives – or ones we are forced to make by our local infrastructure. So there is a huge amount that can be done quickly, and often painlessly. For immediate change here are our top tips:

- Drive less, particularly in cities. Where you have other options walk, cycle and use public transport systems. Many people, especially in urban areas, already live this way. If you are one of the 70 million people expected to buy a brand new car this year, then as soon as you can afford it be part of the tipping point to get all new cars to be EVs and bring an end to new petrol and diesel cars.
- While behind the wheel, we can save up to a quarter of our emissions from exhaust pipes and eroding tyres by chilling out our driving style. Slow down, keep a steady speed and maintain tyre pressure. This will also reduce friction between tyre and road, so fewer particles fly off tyres and brake pads and into kerbside lungs.
- Burning coal or wood within urban areas should be avoided in wealthy countries, and we should help poorer countries to phase these out as quickly as possible.

TEN-YEAR TARGETS

Over a decade many new innovations and initiatives will come about and one of the best things we will each be able to do is to give them our support and encouragement. But based on what we know today here are a few ideas for ten-year goals.

- While many of our Earthshots involve turning things around in the next ten years, cleaning our air justifies aiming for absolute zero. Ask yourself if your Earthshot could be one of these: 'By 2030, I will no longer directly contribute to air pollution' or 'By 2030 every part of my town will have safe air quality.' You could focus on bringing your professional skills to the table: to help those in your country forced to breathe the worst air, as a lawyer you could bring legal cases; as a health-care worker, you could highlight the medical implications of bad air; as a data expert, you could map trends; as an engineer you could invent new technologies.

- In cities with good public transport systems and bikes and scooters for hire, a car may be entirely unnecessary, even for a busy lifestyle. For many people, the options for working from home have ballooned since the pandemic lockdown. Take advantage, to avoid both wasted time and the pollution from the commute especially if your home would be in use by your family during the day anyhow. Walk to school. Maybe an e-bike or scooter would do the trick. If you are planning on moving, try if you can to move to somewhere you won't be overly dependent on motorised transport.

- If you can, make your current car your last one. But if you need your own motorised transport, go electric as soon as you can afford to do so. Do it soon, and you will be an early adopter and helping to hasten the point they become normal. As that tipping point approaches innovations in charging infrastructure and manufacture will accelerate, EV prices will come down and the system will have changed.

- If you have a garden, give it a clean-air makeover. Compost rather than burn garden waste; it is better for your garden soil and the air. Make your garden organic, by ceasing to spray pesticides. And plant trees and other vegetation to soak up pollution and quieten noise. The greenery will also cool the air, shade buildings, calm nerves and encourage nature.

COLLECTIVE ACTION

Some of the most exciting changes over a decade will come from communities, businesses and even whole countries changing. Often these changes begin with a few individuals making a start and inspiring others.

- If your personal actions have allowed you to start cycling to school or work, how can you make a bigger difference? The obvious answer is to encourage others to do the same. If there is a cycling club near your home, work or college, then join it. If not, maybe start one. If you are still one of a brave few on two wheels, figure out why that is and how to encourage others. Are there safe places at work or college to leave a bike? Are there showers? If the roads are dangerous, push for bike lanes, or maybe a group ride would help. Maybe a bike-hire service at a local bus or train station would help. The lesson of recent years is that once a tipping point is reached, many people turn to bikes as an easier commute than cars, and one that is cheaper than public transport. Suddenly you are not a bunch of people coming to work or college in Lycra; you are the new normal.

- Urban areas are shared environments where we all have a right to our say about what is and is not allowed in our neighbourhoods. So be vocal. Be political. Be local. Do not wait for the electric vehicle revolution. Get cleaner air in your streets by campaigning for low emission zones, traffic-free streets and bans on large dirty commercial vehicles. Push for better public transport, and then support it. The old slogan 'use it, or lose it' applies nowhere more clearly.

- We can counter the impact of air pollution by campaigning for more street trees and to protect parkland and rewild unused land. Push the owners of public buildings, offices and factories to set aside land for vegetation. Support those that are making improvements.

- Outside urban areas, ending reliance on a car may still be impractical. Those who do not have access can miss out on jobs and a social life. Maybe you are one of them. Our world is still too often built to meet the needs of drivers. That damages our social fabric as well as our air. Find ways to defeat the hegemony of the car and bring about a local revolution in clean transport of whatever sort works for your community.

- This is a long-term project, and it requires collective action. Reach out to find people who share your ambition. But for those of us who can't find them, we can take the first step ourselves; others will join as our journey becomes more visible. Organise car sharing online. Many villages have their own minibuses to take residents to the nearest town for shopping or even for a night out. Be a passenger. If your community doesn't have one, you could call a meeting to set one up; be an organiser. But also lobby local authorities and transport companies to resume public bus and dial-a-ride services. If a railway runs through your area, campaign for a station. If there is an old route, maybe it can be revived. Such campaigns can take years. Give yourself a decade to make it happen.

• Even in advanced developed nations with stringent environmental laws, there are still dangerous industrial emissions that are worth campaigning to get shut down, to protect everyone's lungs. Coal-fired power stations should certainly be shut, on any number of grounds. Likewise, some old waste incinerators, ageing steel and cement works and landfills. Many farms need to clean up, too. Livestock emissions of ammonia can add dangerously to smogs. If you are in the wrong place, shipping and aircraft emit dangerous amounts of smoke, particulates and sulphur and nitrous oxides. Other menaces are rogue emissions from factories or burning tyres – often at a time when they think regulators are off duty. Be vigilant and call them out.

The bottom line is that clean air and transport systems that do not cost the Earth (financially or ecologically) should be universal rights. Campaign for local, regional and national governments to deliver them. But also be sure to support them and the new initiatives that come along. When there is a clean option, get on board.

Climate

More clearly than with of the other Earthshots, we know where we aim to get to on climate, and the time frame necessary. Our vision is a necessity: to re-establish a stable global climate system. And to do it in time to prevent the rising concentrations of heat-trapping gases from pushing the climate, ice caps and rising tides beyond the point of no return.

Climate scientists are clear that in order to achieve that stable climate system, we need to halt global warming at no more than 1.5 degrees Celsius. We are already above 1 degree. We know this requires ending human-caused additions to greenhouse gases in the atmosphere – especially of carbon dioxide, which once there sticks around for centuries. The first, essential step is to halve emissions in the coming decade.

So our vision has to be of a world that has kicked its addiction to fossil fuels. A world in which coal burning is virtually banished by 2030, and oil and natural gas are in sharp decline. A world in which tropical deforestation has ceased, and the hesitant return of trees and other carbon-absorbing ecosystems now seen in temperate latitudes becomes part of a global programme to reforest the planet and restore nature.

Fixing the climate won't fix everything else. But it will make the other Earthshots many times easier, whether that is tackling dying oceans, smoggy air, food insecurity, accumulating waste or collapsing ecosystems.

FIRST UP

There is so much to do to curb our carbon emissions. And while we know that governments and businesses must lead, many people want to know what simple steps they could take to play their part if they're able. The good news is that there is a lot we can do quickly around the house. As with all of these tips what we can do greatly depends on each of our economic situations. Only the wealthiest among us will be able to perhaps buy a new carbon saving technology to support it in its early adoption but most of us can save both money and carbon at home.

- Choose a low-carbon energy provider. If you live somewhere where electricity supply companies compete to provide power and advertise their carbon footprints, or the mix of sources of their power, then make the right choice for your carbon footprint. That could more than halve the carbon footprint of your electricity by taking you from a coal- or gas-burning supplier to one tapping into solar panels or wind turbines. Still, no country has zero-carbon energy, so we also need to reduce our energy consumption.
- For many people, up to a quarter of carbon emissions

come from the home, so give yours an energy makeover. We need to make sure we have energy-efficient light bulbs, and that appliances such as refrigerators, boilers and washing machines have a good energy rating. If not, that needs fixing at the next purchase. But don't rush to the store, because there is a big carbon footprint in manufacturing those goods that negates the carbon saving.

Some tech is good for our carbon footprint. Laptops are designed to be energy efficient, unlike PCs. Microwaves use less energy than hobs or ovens. Dishwashers are often OK, because they use less hot water than most of us get through in a regular wash-up in a basin. But big-screen televisions and tumble dryers are energy guzzlers. Don't leave electronic equipment on standby: doing so can use a staggering 10 per cent of a typical home's electricity.

- Turn down the washing temperature. For most clothes 30 degrees will work as well as 40 degrees, and will cut a big chunk of the household water-heating bill and carbon footprint right there.
- Do not heat or cool your home unnecessarily. In a northern European country, even a one-degree turn-down can take a quarter of a tonne off your household's carbon footprint. Practise 'passive' temperature control. Retain heat by putting up lined curtains and keep out unwanted sunlight by drawing the curtains or closing the shutters when the sun gets hot. At night, open the windows for ventilation.

Another route is to rethink our carbon footprint outside the home. Act urgently on what is (except for frequent fliers) probably your biggest footprint: driving. We have already discussed this in the section on air pollution. A detailed study of travel by people in seven European cities found that switching to a bike for just one significant journey a day can reduce your carbon footprint by half a tonne over the course of a year – the equivalent of a one-way flight from London to New York.

- Cycling and walking are better for your health, your state of mind and your carbon footprint. If you need to drive then carshare wherever possible and if you can afford it switching to an electric vehicle should substantially reduce your carbon footprint
- Flying is a harder one – the shift to very low carbon aviation is going to take a lot longer than with cars so while we need to support the new innovations they will take a while before the climate benefits. For many of us in wealthier countries where flying is frequent there's no question it's our biggest contribution to climate change. Covid forced most of us to be much more grounded. But as the world opens up again, we don't have to return to the skies in the same way. In particular lockdown has shown that in the digital age much business travel is unnecessary and many companies and employees alike found they have welcomed the change. Travel and the cultural understanding it can bring is important, especially in a world facing challenges we must address as a global species. We

just need to think carefully about our purpose each time we fly and its impact on our planet.

TEN-YEAR TARGETS

Climate change impacts every aspect of our lives, but that means we can find a climate Earthshot almost anywhere that motivates us. Our most obvious – and necessary – ten-year target would be to follow the call of scientists for the world as a whole, and halve the carbon emissions of you and your household this decade. But you could support the wider shifts in society as well. For instance, you could decide to 'reduce business flying at my company by 80 per cent'. Or you could grow your good footprint by deciding to 'change my career to a field that helps tackle climate change', or 'establish a community renewable energy scheme that benefits those in fuel poverty'.

- If you are among those fortunate enough to have a pension or other form of saving, there's an often overlooked way to both reduce your own impact and start to change the system. Such savings are invested on our behalf by a bank or pension fund. Total pension assets alone worldwide amount to $32 trillion. What if over the next few years we all moved our savings and pensions out of those things causing damage and into investments we would be proud to have our money in? Moving your pension can sometimes require your whole company moving its fund and will inevitably take longer and require support from colleagues. But in taking these steps we can do more than just feel good about our savings, we can send a powerful economic signal to politicians and businesses about what we hold to be valuable, and give a very real incentive to change where the big money goes.

This of course doesn't only apply to climate change – increasingly there are options to move your money out of all kinds of investments that leave a negative footprint and into ones that bring a positive impact.

- If you have control of the structure of your home and savings to invest in your property you could take on some bigger works than those in the earlier section. Firstly make sure it is properly insulated: not just the roof, but any cavity walls, and double-glaze windows. Insulation keeps you warm in winter, and will keep the heat out in summer, too. To further reduce the need for air conditioning in hotter climes, you may be able to paint roofs and walls white to reflect the rays. Or put up solar panels: as well as generating electricity, they will stop the sun heating the house. Shade outside walls with plants. After a few years' growth, a tree or two works wonders.

- You may be able to enhance control over your footprint further by installing your own energy source. Most likely that would be by letting a company put solar panels on your roof. You can use the energy from the panels to reduce the power you take from the grid. On sunny days, you might even end up selling solar power to the grid. Check out the options where you live and you can aim to take them up as soon as you can afford it.

- Get ready for heat pumps. They are expensive right now but if you are able to afford them they can make a huge difference – and as more people install them the prices will come down. Right now, many of us have little choice when it comes to how to heat our homes. In much of Europe, natural gas is the norm. Heating a typical family home that way may emit more than 2 tonnes of carbon dioxide in a year, making up as much as a fifth of your carbon footprint. But electric heat pumps are coming. They work like air conditioners in reverse, sucking warmth from the air outside, or from the ground, and pumping it to radiators indoors. If the electricity comes from low-carbon sourcing, the emissions from heating the house can be halved. Supporting those that are finding affordable ways to bring them to mass markets, from government subsidies to new innovations, will make the roll-out faster.

- What should we aim for on our carbon footprint? We don't suggest a specific figure. It depends too much on how national energy and transportation and food systems are set up. But if you are in a wealthy country you might want to make sure you are below your national average. Then head for half your national average. That would be 3 tonnes of carbon dioxide per year in the UK, 4 tonnes in Germany, and 8 tonnes in Canada, the United States or Australia.

COLLECTIVE ACTION

Some of the things we have listed to do at home are hard to achieve if you are in rented accommodation or in an apartment block, because they involve changes to the fabric of the building. But if you have a tenants', students' or residents' association, use them to push those in charge of the building for change. Many of the big changes to fix the climate come from working together to come up with new ways of doing things. Here are a few ideas to get you started:

- Get involved in making your town or city a beacon of low-carbon living. That will require bringing together local groups, residents' associations, politicians, businesses, consumer organisations, trades unions and others. You will probably find many like-minded people, often awaiting ideas and leadership, or who want to join in. Actions that can be done at city-wide level include controlling traffic to reduce vehicle emissions, installing recharging networks for electric vehicles, promoting heat pumps and domestic insulation, encouraging retailers to lower the emissions 'embedded' in the produce they sell, reducing emissions from bus fleets, painting roofs white to reflect sunlight, making cycling safer, and managing public transport better.
- Join networks of like-minded citizens round the world to share ideas and encourage officials to see their efforts in a wider content. From Copenhagen, Denmark, to Aberdare, south Wales, and New York

City to Curitiba, Brazil, and Singapore, there are good examples for us to follow.

- Use the power of money. Divestment campaigns have put pressure on large corporations. You may be a direct shareholder, or more likely have a pension with investments in big oil companies, for instance. Or, as a student at a university or an employee of a public body, you can persuade your institution to pull its money out of fossil fuels. Join others to get organised, and put forward motions at annual general meetings.

  An alternative route is to use shareholdings to change policies. Environmental activists have found that mainstream shareholders such as asset managers and banks often feel that companies are being too slow to respond to climate change. Shareholders in the United States recently voted to require the boards of ExxonMobil and Chevron to make them viable in a world of decarbonisation. High-profile campaigns aimed at ending sponsorship by fossil fuel companies of museums, sports events, arts venues and other activities can also grab attention. But choose your targets carefully. Work out who your friends are, as well as your enemies.

- Support the solutions. Whether it's a new low carbon supplier to your company, a new sustainably sourced food shop in town, a great idea at school – new ideas to fix our climate don't just need inventing, they need supporting.

- Fighting climate change requires national and international political will, too. Hold leaders and big corporations to account. In this respect, the law can be on your side. You might want to get involved in environment groups taking governments and corporations to court. In 2021 environment groups won a Supreme Court ruling in Germany requiring the government to do more to reduce national emissions in the coming decade, because not doing so would put too great a burden on future generations. And Dutch groups won a case against oil giant Shell requiring it to accept the same emissions reductions targets as European governments – a 45 per cent reduction in emissions by 2030 – on the grounds that there is a 'broad international consensus about the need for non-state action, because states cannot tackle the climate issue on their own'.

Nature

After spending the twentieth century wrecking natural ecosystems, we now have the challenge of a global restoration of nature. The United Nations has declared the 2020s to be a decade of ecological restoration, and everyone can join in. More natural forests, more urban trees, more grasslands and wetlands, more tundra peat – all have their role in stemming the catastrophic loss of biodiversity, improving nature's ability to capture the carbon dioxide we have unleashed into the atmosphere, soaking up pollution, making rain and cooling the air.

Sometimes we will want to share land with nature,

bringing back hedgerows or planting more trees in our fields and cities. Sometimes we will want to make room for nature by stepping back – by doing agriculture better, or finding ways to grow and make food without using conventional farms at all. However we approach it, we will want to reduce the planet's domesticated livestock, giving grasslands back to wild animals while reducing the need to grow crops to feed the former.

Sometimes we think of deforestation as something being done by other people. However, these days most forests are burned or chopped down to grow commodities for sale to consumers, mostly in richer countries. An average Western citizen with a typical diet is responsible for the felling of four trees each year in the tropics. That may not sound a lot, but the European Union reckons that its half-billion citizens are responsible for almost a sixth of all tropical deforestation. Palm oil from former forest land in Indonesia is used in one in three of all supermarket products; soy beans from the Amazon feed Europe's chickens; then there is beef from Argentina, rubber from Vietnam, chocolate from Ghana, and so on.

When it comes to restoring nature, there are lots of ways to reduce your bad footprint and increase your good one. Most come down to making space for nature and reinventing our diets.

FIRST UP

Around the world we all eat very different diets based on factors ranging from culture to wealth to geography to personal taste. There's no perfect guide to a sustainable diet for each person but here are some good places to start.

- We all have to eat, but when we buy food or other products, they could have been grown on former rainforest. Supply chains are often opaque, so making our way down the supermarket aisle figuring out which purchase is best, or whether they are all as bad as each other, is hard work. Do better by looking for products with recognised 'sustainability' labels. Among them are the Roundtable on Sustainable Palm Oil, set up by WWF with some big food businesses; the Soil Association for organic produce; and the Rainforest Alliance, which certifies everything from coffee to bananas and house plants to nappies. There are also Forest, Marine and Aquaculture Stewardship Councils, again initiated by WWF with support from parts of the relevant industries, and the Fairtrade label, which is specifically directed at raising standards for farmers.

  None of these labels is perfect. Each hides internal battles, usually between those who want to recognise only the best and others intent on driving better standards for the whole industry. But still, they are a decent benchmark.

  As new technologies like blockchain get used in commodity supply chains it should become easier to trace products and the companies that become really transparent and offer truly sustainable food will need and deserve backing from customers

- Eat more plants. No one likes to be told what to eat

but if you are concerned about your footprint then eating more plants is generally the way to go. You are unlikely to find sustainability labels attached to mainstream meat products because – whether you are looking at the deforestation footprint, the water footprint, the carbon footprint, the nitrogen pollution of rivers, or almost any other measure – most meat does not look good. Especially red meat. By some estimates, up to three-quarters of agricultural land round the world is devoted to grazing livestock or growing the crops to feed them. It takes a lot of land, because getting our food by growing crops to feed to animals that in turn feed us is hugely inefficient.

So, in the first instance, eating more plants and cutting down consumption of meat will mean massively reducing the amount of crops that have to be grown to feed you. Because it takes 2 kilograms of feed to produce a single kilogram of chicken, 4 kilograms for pork and 8 kilograms for beef. There are similar statistics for water use, and for nitrogen pollution. Livestock that graze on pastures are better, but even they have a big footprint because of the land they take – and the methane that they produce. It is certainly possible to find meat that comes from well managed farms that do a great job for biodiversity – but there's nowhere near enough of that to meet current consumption levels.

If you are an average European, getting through 80 kilograms of meat a year, cutting that out could reduce the land needed to feed you by around 50 per cent. An average American eating 110 kilograms could make an even bigger contribution. If you aren't ready to stop eating meat altogether, then try eating meat only every other day. Even in the gourmet's heartland of France, meat is off the menu at least one day a week in schools, as part of a zero-emissions government strategy.

• Buy local produce if grown sustainably. It is unlikely to have destroyed any forests, and transport and wastage rates are lower. The benefits are greatest if the produce is fresh, however.

Oh, and don't buy mineral water, if your local tap water is safe to drink. It's not the water in the bottle that should worry you, but the 7 litres of water needed to make the bottle, and the massive waste of carbon dioxide in shipping it to you. Bringing bottled water from Scotland to England is bad enough, from France it is nuts, and from Fiji, a distance of 15,000 kilometres to the UK, it defies rational explanation. Refill a water bottle from the tap when travelling, and insist on tap water in restaurants.

TEN-YEAR TARGETS

Restoring nature offers many Earthshot opportunities. You could reduce your bad footprint by deciding to 'halve the environmental impact of my food choices' or even widen that to include all your consumption habits and 'become deforestation free within a decade'. Or you could target boosting your good footprint, by finding ways to 'make my garden as good for wildlife as it can be', or 'double the number of trees in my town', or even 'join with others to find land in the countryside for a wildwood project'. Remember, none of this is a substitute for holding our leaders to account, but these are suggestions for everyone who is eager to play their part.

- Share space with nature. If you have a garden, do it there. Give up the lawn and let nature reclaim it. Put in a pond, and you will be amazed what shows up, from birds to amphibians. The great thing about rewilding is it can be as small or as big as you like. Even putting a window box on a sill in the sun will attract insects, adding to urban wildlife. A few people can be more ambitious on their land. If you have a farm, go for it. Set aside parts for natural woodland or simply for the grass to grow. Or go further. Earlier, we looked at what Isabella Tree did at Knepp in Sussex, rewilding an entire farm. If you do, you could have tourists and naturalists queueing up to visit. If you have some spare money, you could buy a wood, or buy a field and make it into a wild-flower meadow.
- Going vegetarian even a few times a week is good, but reducing your consumption of dairy products will also have a big impact. This is because eggs and dairy products such as milk and cheese come from livestock, though their footprints are dramatically better than meat. According to a detailed dietary study by the University of Oxford, going vegan would reduce both your carbon and water footprints from food by 60 to

80 per cent. So even a few plant based meals will start to make a big difference. Perhaps more importantly you will be part of a growing social movement. And as more and more people switch to a more plant-based diet, more restaurants, cafes and supermarkets are offering more plant-based options. A lot of food labels now indicate whether they are vegan compliant – encouraging ever more people to make the change. Give it a try.

• While most of our consumption footprint on nature comes from food, we need to think about the sustainability of other land-based materials that we consume. Wood-based products such as furniture and paper should either be from recycled sources or carry a certification label such as that from the Forest Stewardship Council (FSC). Always look before buying. Even books usually carry the FSC logo. The labels are not foolproof, but they are a good sign. Supporting people who make the effort to source their materials carefully rewards them and makes that practice more likely to spread.

Sustainability also comes from using less. Buying furniture and clothes that will last. If a bag can be for life, so can a table. Also look for clothing companies that try to source cotton sustainably, and recycle textiles. The WWF-backed Better Cotton Initiative claims that a quarter of the world's cotton is now grown 'more sustainably'. If you find a company or shop that supplies good-quality clothes you like in a sustainable way, share that with friends and beyond via social media. Join the movement to replace instant disposable fashion with sustainability and quality.

COLLECTIVE ACTION

Restoring your locality is a great collective endeavour that spreads the word and invites others to get together to do the same. Find like-minded people – maybe gardeners at local allotments or parents whose kids use a local park – and get campaigning for local rewilding projects on public or unused land somewhere near you. Get a campaign going to persuade your council to rewild part of a local park, an old railway embankment or abandoned scrubland down by the sewage works. Nature loves old industrial land too, as well as cemeteries and the banks of rivers and canals. If your local council won't do it, then perhaps they will give permission for a local group to take matters into their own hands.

Urbanites might want to join the ranks of guerrilla gardeners, staking out urban wasteland for nature. Some guerrillas grow crops; some are into rewilding. Make the most of it. Bringing nature back to our cities benefits everyone.

- In the countryside the possibilities are even greater. Rewilding Europe is a growing movement, and there are similar initiatives springing up in other parts of the world. Many of the most successful projects are organised by rural communities. To take two examples, near Aberystwyth, in Wales, they are creating the Cambrian Wildwood, a mix of bog, heath, rough pasture and forest covering 300 hectares so far, and with plans to expand to ten times that area; and in the Moffat Hills of southern Scotland, local enthusiasts are behind the Carrifran Wildwood, sprouting across 2,500 hectares of rewilded valley. The latter created the Borders Forest Trust to buy up little used farmland in a valley and restore the wild woods that occupied the land there some six thousand years ago: there should, they agreed, be 'access to all'. They soon found that hundreds of volunteers joined the endeavour, planting seeds of native trees, protecting the saplings and enjoying the new wildwood as it grew.

Oceans

Healthy oceans once again full of fish, nurtured by resurgent coastal ecosystems such as coral reefs and seagrass, are essential to our vision of a restored planet. The oceans cover two-thirds of our planet. They are vital to biodiversity, to our weather and to the water, carbon and other systems that regulate our world. They make the land habitable for humans and everything else. They are also spectacularly resilient. As with those on land, marine ecosystems can recover fast when the pressure of human activity is relieved.

Our vision of restored oceans rests on the experience of marine protected areas that have proved to be spectacular successes in places, restoring local marine ecosystems and acting as a hub to seed recovery much more widely. We know too that climate change and pollution from the land pose grave threats to those powers of recovery. But if we turn things around now, we can continue to fish the oceans and benefit from their healing powers.

FIRST UP

For those of us who eat fish and plan to continue to do so, then we should support good practices. Find fish you are confident can be sustainably caught or farmed. Eating fish can seem like asking for ecological trouble. It may be caught in the wild by unpoliced trawlers, or raised on fish farms that pollute local waters and may depend on wild-caught fish for feed. But think again. For millions of people fish are an essential part of their diet and livelihood, so we need to find ways to support those that do it right, and boycott those that don't. As we saw earlier, there are signs of improvement. Eating sustainably caught or farmed fish is a much more climate- and land-friendly way of getting animal protein than eating meat. Wild fish find their own food, and farmed fish such as salmon convert virtually all their feed into protein.

- Buy locally caught fish where possible if it can be shown to be sustainable. Local fishers may not always be flawless, but they are usually more artisan than industrialist, and they have a direct interest in maintaining stocks. Regulation is also generally tighter in the coastal waters they harvest than it is in more distant waters, where giant super-trawlers may rule. You might also want to support local businesses. Otherwise, there are labelling schemes for sustainability. The Marine and Aquaculture Stewardship Councils, both initiated by WWF with food industry backing, have been criticised, but they remain useful barometers.

- Before ordering fish in a restaurant or at a fish counter, ask questions about where it came from and whether it was sustainably caught or raised, and be satisfied by the answers. If not, don't buy. Restaurants and retailers should provide such information routinely. Some specialists advertise their virtue. But others think we don't care. So, whether it is the restaurant or fish stall, the supermarket or the cafe, always ask. So they know you do care. It all adds up. The pressure builds, the good guys are rewarded and sustainable fishing can become the norm.

- Don't let waste end up in the ocean. One quick win that we can all achieve today is to make sure that dangerous chemicals from our electronic waste don't end up in the ocean. If thrown into landfill, hazardous chemicals can leach into the soil and be washed into rivers and ultimately the ocean. So always properly recycle electronics.

TEN-YEAR TARGETS

Helping the oceans is difficult. Most of us have little direct connection with them in our daily lives, so making an impact seems hard, or at any rate remote. But a good Earthshot for anyone would be to ensure you do no harm, by deciding to 'ban single-use plastic from my life' and 'work to eliminate it from my workplace and my community', and for fish eaters to 'eat fish only from sources that do not harm the ocean' or more ambitiously to 'work to eliminate unsustainable fish from my local shops and restaurants'. If you live in or near coastal communities, get stuck in locally. Campaign for marine protected areas, and to protect shorelines from development. Publicise the importance of maintaining natural ecosystems, whether coral reefs, kelp forests or seagrasses. Be an educator. Many people simply do not know about the ecological riches beneath the waves.

- Work to eliminate single-use plastic from your life as much as possible. Currently we recycle just a small fraction of our plastics. Plastic cutlery and straws, lids and wrappers, bottles and bags; almost all the plastic ever produced in the hundred years since it was invented remains out there somewhere.

  Until surprisingly recently, plastic waste was largely ignored by environmental scientists because it was regarded as being chemically inert. That made it unsightly, but not of great interest or importance otherwise. We now know that was a big mistake. Because being chemically largely inert means plastic waste doesn't go away. It accumulates, ensnaring wildlife in the oceans, and gradually breaking up into microplastics and then even smaller nanoplastics, which may be doing untold harm simply by getting buried deep in lungs, guts and other organs.

  And every year humanity adds another 335 million tonnes – roughly the same as the weight as humanity itself. So most of us are adding approximately our own weight in plastic to the world's waste pile each year.

• If you plan on visiting coastal waters while on holiday, treat them with respect. Support local initiatives that protect wildlife and keep local waters clean. Whale watching can be fun, but don't go if you suspect the boats are harassing the animals. Make sure you don't sunbathe, or do anything else, on a beach where turtles nest. This is not just because you may damage the nests. Even the lights from distant beach bars will disorientate newly hatched turtles trying to steer by the moon to get back to the ocean.

Diving, especially around coral reefs, is a great adventure. For many coastal communities, taking people to the reefs – and feeding and watering them afterwards – provides an attractive alternative to fishing, and is a big benefit to local fish stocks and marine life. We saw that with the Castro family on the beach in the Gulf of Cortez. But, as with land-based eco-tourism, there are dangers in numbers. Many reefs are damaged by too many people, especially those inexperienced at how to avoid harming them, or who think it is OK to take a bit of coral home. So beware. Do research in advance to make sure you will not be diving into an ecological disaster area.

COLLECTIVE ACTION

Nobody owns the oceans, so we have to step up and make a difference by campaigning for proper marine protection.

As we have seen, one thing that has been shown to work is the creation of marine protected areas. Safeguarding even quite small areas of coastal sea can within a few years result in the spectacular recovery of marine life. To transform the oceans, and harness nature's prodigious powers of recovery, we need many more of them, with better protection. Marine scientists call for a third of the world's oceans to be protected. We have just a fraction of that now. But by organising globally – and keeping faith with local fishing communities, who may fear no-catch zones but will also know best the potential benefits – protection could scale up fast, delivering healthy oceans and more fish for sustainable harvesting.

- Do regular beach cleans, if and when you can, and join beach-cleaning gangs. For those who live near the coast or visit it regularly, a good move is to informally adopt a stretch and clean it every time you pass. It's a never-ending job, and it won't rid the oceans of plastic pollution, but it will help some marine life. Most of us who don't get to the coast much probably still live close to a river. Rivers are all too often conveyor belts taking rubbish to the sea. So do a regular clean in a river and on its banks. Use your example to inspire others. In the Caribbean, teacher Kristal Ambrose became known as Kristal Ocean for her battle organising youth camps to clear up plastic waste from the beaches near her home in the Bahamas. She formed the Bahamas Plastic Movement, pestering ministers until they passed a law banning all single-use plastic on the islands. You might want to follow her example.

But remember that while human detritus is not just unsightly but also dangerous for wildlife, a certain amount of flotsam and jetsam is normal, and of ecological value. Mud and beach creatures find food and shelter amid the beached wood and seaweed. A shoreline cleaned of such items in order to improve the experience of sunbathers can be a biological desert. So leave behind nature's detritus.

- If you live somewhere this is happening, join efforts to rewild the oceans. We saw in Part 2 how marine biologist Bob Orth has been replanting seagrasses on the east coast of the United States. He didn't do it all himself. Hundreds of volunteers, many of them students, did most of the work. Similar work elsewhere has restored coastal kelp forests, mangroves and even coral reefs. If you are a marine biologist, don't just study, restore. Assemble gangs to do the work. If you live or study anywhere this work is going on, join in. All you probably need is a pair of flippers and some free afternoons.

Waste

Waste is the root cause of most of the planet's problems. Waste gases in the air above our streets and in our global atmosphere; waste plastic in the oceans; liquid industrial waste and agricultural nitrogen in rivers; wasted food; waste construction rubble and mining ores; waste pesticides accumulating in the Arctic; the waste products of our own bodies. The very word is a problem. Almost all this 'waste' could be turned into raw materials and valuable nutrients in a circular economy. Our vision is of a world where this circularity in our use of resources happens for most wastes most of the time. In such a world, we will 'mine' the trash of past eras, cleaning up the planet as we go. We will wonder at the folly of ancestors who threw away stuff so wantonly. Along with stabilising our climate, cleaning urban air and beginning the great restoration of nature, creating a circular economy is the great challenge of the twenty-first century. If we want our children to have lives worth living, it must start here and now.

FIRST UP

Our top priority should be putting an end to food waste. Doing so will save money and time as well as reducing every footprint and keeping your rubbish bin as empty as possible. Food that has been cooked for our meals can become creative leftovers: reheat and refashion, fill a sandwich for lunch tomorrow, make some soup. Check out cookery writers with great ideas. As a last resort, use it as compost in your garden or window box. If you can't compost, many local authorities now collect food waste: if yours doesn't, ask them to. Waste nothing: literally.

It is a scandal that a third of all food – grown at such great expense to forests and grasslands, rivers and oceans, and the atmosphere, not least through methane emissions from landfills – is simply thrown away. It works out at roughly 100 kilograms wasted by average citizens of developed nations every year. Or around one meal every day.

Not all of this is done personally. Much of it disappears into bins at the back of supermarkets, restaurants, canteens and warehouses. Supermarkets routinely throw away up to a third of the fresh fruit and veg on their shelves. WRAP found that almost a fifth of lettuces grown in Britain never get beyond the farm gate, along with a tenth of strawberries. But, whoever is responsible, it is a huge waste of all the land and water, energy and farm chemicals and time required to produce it in the first place.

Try a survey for a week: what causes food waste in your household? Is it impulse buys? Buying too much? Being tempted by offers such as 'buy one, get one free'? Is it putting too much on the plate? Whatever the cause, find the best way of not repeating it. Always ask before buying: do I actually need this? Carry and store fresh food carefully, so it is less likely to bruise or rot. Check out places reducing waste by selling oddly shaped vegetables rejected by the big supermarket chains.

Two-thirds of what we throw away remains edible. Ignore 'best before' dates, and think carefully about whether to abide by 'use by' labels if the food appears still good to eat. If you are not going to eat something that is in a good state, make sure it gets to a food bank or somewhere else where they need it.

- At restaurants, ask for a doggy bag to take home any food that doesn't get eaten. In France, that is a legal requirement – and in big-portion America, extremely common. France, incidentally, has one of the lowest food wastage rates in the developed world. Supermarkets there are required to distribute any food they have left to food banks and charities. There are prison penalties for those that fail to do so. Maybe, as a gourmet nation, they value their food more.

- Target packaging, especially. It probably makes up nearly half your waste bin. Clearly we should be recycling much more. As we have seen, the technologies to do this are often still work in progress. But by far the best solution, as with most waste, is not to make, buy or use it in the first place. So, buy loose goods and support the stores that offer them. Carry a water bottle. Take your own mug for a coffee refill. Turn down packaging when offered. Use soap and shampoo bars rather than bottled liquids. Always use paper bags rather than plastic, and 'bags for life' rather than single use. Support bottle return schemes where they exist – Germany, Norway and Sweden recycle 90 per cent of plastic bottles this way. If the service is available, get milk delivered in glass bottles. You won't be 100 per cent successful, but you can get close.
- Think before flushing. Don't flush cotton buds or wet wipes down the toilet. They won't be removed by most sewage works and will probably make their way to the ocean, or on to the beach. Some pharmaceuticals flushed away are turning up in rivers at concentrations that may harm fish. So take unused medicines back to the pharmacy.

TEN-YEAR TARGETS

To eliminate waste, ask yourself what scale you would like to work at? One choice is to find ways to 'reuse or recycle everything in my household', or to 'always buy things that will last'. But you could think bigger. If you are a good organiser, then decide to 'help my community halve its waste this decade'. If you are good at understanding processes and solving technical problems, work to 'eliminate waste from my company's supply chain.' Or imagine yourself the new Tom Szaky who, as we saw earlier, turned concern about food waste on his university campus into an international business rethinking the supply chains of the world's biggest consumer companies. Waste may sound dull, but fixing it is fundamental to fixing our world.

- As a longer-term goal, we need to think about our stuff – the equipment and devices, clothes and toys that we surround ourselves with. Okay, buy less of this stuff. But once we have that stuff, what should we do? Before getting rid of anything, even for recycling, imagine for a moment the materials required to make it. Think of it not as a footprint but as a materials rucksack. A heavy one. Your mobile phone may look a small thing, but it has a rucksack of materials used in making it that weighs about 75 kilograms – more than a thousand times its own weight. A PC computer has a one-tonne rucksack. As we extract metals, rare earths and construction materials from the earth, we leave behind a vast trail of mining and refining waste.

Typically, a single kilogram of metal requires hundreds of kilograms of ore from mines. All told, a typical European has a materials rucksack that grows by about 50 tonnes a year. By the end of your life that is something like 4,000 tonnes. Over the next decade there will be many different innovations to reduce that but they will need encouragement and support. From companies that contract you electronics rather than selling you them so they fix and reuse the parts; to tech that's built to be repaired rather than discarded; to new ways of repurposing and recycling materials. There are thousands of new ideas to be launched on the road to turning waste into resources. Could one of them be yours? Throw as little away as possible. Be a reuser and repairer of anything and everything. Find new uses for stuff. Even trivial items, especially plastic ones that won't biodegrade. Toothbrushes make great scrubbers for difficult corners in the kitchen or bathroom; plastic containers make good plant pots or bird feeders. Once you start thinking about it, you will come up with many ideas.

If that fails, many technical colleges and night schools take equipment such as bicycles to repair as part of their courses. Check out websites that arrange swaps or will just pass on stuff for free to someone who needs it. A child's playthings often outlast their interest, for instance. When that doesn't work, recycle. Always make use of charity bins that collect clothes, shoes and other such things. They may find a new home, and at worst will be reused as rags or safely disposed of. Remember that many clothes are made with some artificial fibres, which often means plastics. Like food packaging, your body packaging won't biodegrade and may end up dumping microplastics into the ocean.

- In many countries there is legislation that requires electronic waste to be collected separately, so the materials can be reclaimed and reused, safely and sustainably. Make sure to do that. Batteries should be recycled too, along with printer cartridges, scrap metal, aerosols, cooking oil and much else. Check your local recycling centre for the full range.

COLLECTIVE ACTION

As the drive to build a waste-free world gathers momentum, innovators and entrepreneurs will spot new business opportunities – from making products that are easy to repair to promising to take back products that are done with, for reuse or recycling into new products. More products will hit the shelves that are part of a closed loop system in which waste from one process becomes a raw material for another. Consumer pressure will drive this, but so will the flair of individuals within companies. Perhaps you will pioneer something in your industry or community, or perhaps you will publicise someone who does. Once it is clear that enough of us want to go beyond a throwaway society, the ideas and innovations will come thick and fast. But we have to reach that moment. We must all do our best to bring it on.

The United Nations and many governments have agreed on a target of halving per capita food waste by 2030. It is a global priority. So ask how your government is doing. And your supermarket and canteen and restaurant. Be part of the solution as a citizen as well as a consumer. Help out at a food bank, collecting or distributing, for instance.

If some categories of waste are not being recycled or collected for reuse in your area, then explore ways of plugging the gap. Send around some social media messages to build up a base of interest and figure out where the gaps lie. Of course you will need to find someone who wants those materials as well, or you just end up with a big pile of junk. Sometimes retail outlets or waste centres will take the hint, find some markets and put in some new collection bins. But also lobby your local council to do the organising. Ask if they have reached targets in Europe and elsewhere for recycling at least half of all household waste. And if not, what their plans are to achieve that. If they have no such plans, they may in breach of the law. And if they have achieved the legal targets, push for them to do better. If San Francisco can recycle 75 per cent of its waste, so can they.

The Power of a Million Earthshots

Achieving our Earthshot challenge to transform the world in the next decade – to create for ourselves and future generations a 'good Anthropocene' – requires leaders and followers, inspirers and geniuses, bosses and workers, campaigners and consumers, voters and shareholders, entrepreneurs and even the reluctant. We can leave no stone unturned, no avenue unexplored and no human uncontacted. We need the urgency of fearing for our future and that of our planet, but also the optimism that comes with knowing what can be achieved.

So, seize the moment. Whoever you are, and wherever you are. It's time to stop thinking about saving the world in terms of what we sacrifice but instead to think about what we can gain.

Certainly reduce your own impact, but also get involved to change the system around you. Support the great ideas and initiatives. Whether as a citizen voicing concerns and offering solutions within democratic systems, or as a member of a church or school, college or club, neighbourhood or city fighting for change. Whether as an investor or pension holder ensuring your money is used for good rather than harm, or as an employee working within your own company or institution to identify how they can reduce their footprints, or by joining one of the increasing number of organisations with a mission to do good. We have ten years to do this.

Your Carbon Footprint in Numbers

We cannot expect individuals to bear the responsibility for fighting climate change, but the more we know about our personal impact the more power we will have in demanding change and identifying the initiatives and new ideas that will make the biggest difference and deserve our support. A big problem when we first try cutting our carbon footprint is knowing what is trivial and what makes a big difference. So let's work our way up from the smallest to the biggest in tenfold leaps. These carbon dioxide numbers are very approximate, but they give you an idea:

1 gram	An internet search
10 grams	One email; one reusable plastic bag
100 grams	A one-kilometre drive in a small petrol-driven car; an hour of video streaming
1 kilogram	A serving of pork (or cheese for vegetarians)
10 kilograms	Four litres of petrol; two minutes of a hair dryer every day for a year
100 kilograms	A hot shower every two days for a year
1 tonne	A return transatlantic flight

Your Water Footprint in Numbers

It is amazing how much water it takes to get us through a single day. Consider these water footprints:

1 litre	A dripping tap in an hour, at two drips a second
10 litres	A typical toilet flush; a minute in a shower; the irrigation needs of a single almond or pistachio nut
100 litres	One person's daily domestic water needs for washing, flushing and the rest; irrigating enough feed for one chicken to lay one egg; growing one portion of rice or wheat for the bread in one sandwich, or one mug of coffee
1,000 litres (1 tonne)	A glass of milk
10,000 litres (10 tonnes)	Irrigating enough grain for enough cattle for one beef steak; a half-kilo jar of coffee; cotton for a single pair of jeans
100,000 litres (100 tonnes)	Enough bread for a year's supply of lunchtime sandwiches
1 million litres (1,000 tonnes)	Roughly your total water footprint in a year

THE
EARTHSHOT
PRIZE

About us

T he Earthshot Prize is a new global prize for the environment designed to incentivise change and help to repair our planet over the next ten years.

The science tells us this is the decisive decade for our planet. While the urgency of environmental issues is felt by many, it is hard to ignore the pessimism and despondency which surrounds them. To overcome this great challenge, this equation needs to be changed. Pessimism and despondency must be turned into optimism and action.

In partnership with world-leading philanthropists, environmental NGOs and high-profile leaders committed to environmental action, The Earthshot Prize was designed and globally launched by Prince William and The Royal Foundation.

The Prize has taken inspiration from President John F. Kennedy's 'Moonshot' which united millions of people around an organising goal to put man on the moon within a decade – a seemingly impossible task but one that was made possible within ten years, highlighting that great things can be achieved when humanity works together.

The Earthshot Prize's challenge to the world is based on

331

five 'Earthshots' – simple but ambitious and universal goals for 2030 – Protect and Restore Nature, Clean Our Air, Revive Our Oceans, Build a Waste-free World, Fix Our Climate – which if achieved will improve life for us all.

Every year between now and 2030, Prince William and a global Prize Council will award The Earthshot Prize to five winners, one per Earthshot, whose solutions make the most progress towards these goals.

What makes the Prize unique is that these winning solutions could come from anywhere; individuals, teams, companies, even cities and countries. And the solutions won't be limited to just science and technology; Prizes could be awarded for leadership or grassroots projects.

The Earthshot Prize is about much more than awarding achievement; it is a decade of action to convene the environmental world with funders, businesses and individuals to maximise impact and take solutions to scale; to celebrate the people and places driving change; and to inspire people around the world that we can work together to repair the planet.

**For more information on The Earthshot Prize,
visit www.earthshotprize.org**

Facebook/Instagram/Twitter – @EarthshotPrize

YouTube – EarthshotPrize

The Earthshot Prize Partners

The Earthshot Prize is supported by its Global Alliance, an unprecedented network of individuals and organisations worldwide which share the ambition of the Prize.

Within the Global Alliance are its strategic funding partners, known as Founding Partners, who are a group of leading global organisations and philanthropists including Aga Khan Development Network, Bloomberg Philanthropies, Breakthrough Energy Foundation, DP World in partnership with Dubai EXPO 2020, Jack Ma Foundation, Marc and Lynne Benioff, Paul G. Allen Family Foundation and Rob & Melani Walton Foundation.

In addition to this, the Global Alliance Partners are non-profit and international organisations committed to the environment and sustainable development, who bring expertise, global reach and serve as nominating organisations each year. This includes C40 Cities, The Commonwealth Secretariat, in respect of the Commonwealth Blue Charter, Conservation International, EARTHDAY.ORG, Fauna & Flora International, Global Citizen, the Green Belt Movement, Greenpeace, John F. Kennedy Library Foundation, National Geographic Society, One Earth, TED Countdown, The Paradise International Foundation, Project Everyone/ World's Largest Lesson, Sustainable Markets Initiative,

United Nations Environment Programme (UNEP), World Economic Forum, WRAP and WWF.

Furthermore, our Global Alliance Members are some of the world's largest and most influential companies and brands who will support our Earthshots, implement ambitious changes within their businesses and accelerate our winning solutions.

The Earthshot Prize is also supported by its Prize Council, a truly global team of influential individuals from a wide range of different sectors, all of whom are committed to championing positive action in the environmental space. Every year from 2021 until 2030, The Earthshot Prize Council will award The Earthshot Prize to five Winners, one per Earthshot. The Members of The Earthshot Prize Council are His Royal Highness Prince William, Her Majesty Queen Rania Al Abdullah, Cate Blanchett, Christiana Figueres, Daniel Alves da Silva, Sir David Attenborough, Ernest Gibson, Hindou Oumarou Ibrahim, Indra Nooyi, Jack Ma, Luisa Neubauer, Naoko Yamazaki, Dr Ngozi Okonjo-Iweala, Shakira Mebarak and Yao Ming.

Acknowledgements

This book encompasses a grand story. It would not have been possible to have told it without the assistance, support and inspiring contributions of many colleagues. The authors would like to thank all those associated with The Earthshot Prize at the Royal Foundation of the Duke and Duchess of Cambridge. In particular Jason Knauf, Amy Pickerill, Rachel Morriarty, Charlotte Pool, Katie Heywood, Rosie Fenning, Edwina Iddles, Claire Straw and Zoe Ware. This team has been challenged with achieving their very own Earthshot in setting up the Prize in the first place, and we have been honoured to accompany them on part of their journey and to have the opportunity to take this important story to a broad readership with the help of their unique platform.

We are very grateful to our mentors Alastair Fothergill and Keith Scholey for their continued guidance and support and to the team at Studio Silverback for their part in helping to enrich and clarify this amazing story. Particular thanks go to the producers, Amy Anderson, Jon Clay, Clare Dornan, Dan Huertas, David Heath and Lizzie Green, the assistant producers, Matt Brierley, Struan Kane, Natasha Filer, Erica Wilson, Claire Sharrock and Sophia Luzac, and the researchers, Jess Hall and Roxy Furman.

We are greatly endebted to the key staff behind the book's production, Nick Davies and Kate Hewson, who have led the charge at John Murray; our scientific advisors, Mark Wright and Owen Gaffney; our colleagues Jane Hamlin, Laura Meacham, Anna Keeling and Rachel James at Silverback and Robert Kirby at United Agents. Special thanks must go to the wonderful, reliable, unflappable Fred Pearce, for helping us pull this volume together in record time and to such a high standard.

Finally, may we thank His Royal Highness, Prince William, for initiating and supporting The Earthshot Prize. We believe it to be one of the most significant undertakings in the history of environmentalism, an incomparable initiative of its time, and a vital energiser for a world that is suddenly aware of the situation it is in, and that must rally quickly to do something about it. His Royal Highness is an inspiration to us all of how to do so.

Colin Butfield and Jonnie Hughes
Bristol, UK
5th August 2021

About the Authors

Colin Butfield is co-founder of Studio Silverback, Executive Producer of the WWF's Our Planet project and an advisor for The Earthshot Prize.

Jonnie Hughes is an award-winning natural history TV producer, co-founder of Studio Silverback, and bestselling author and co-writer of Sir David Attenborough's *A Life on Our Planet*.

Picture Acknowledgements

Images in plate sections:

1. Credit to Mark Fox / Getty
2. © Juozas Cernius / WWF-UK
3. © Emmanuel Rondeau / WWF-US
4. © Marizilda Cruppe / WWF-UK
5. Credit to Kiyoshi Hijiki / Getty
6. © Conor McDonnell
7. © Conor McDonnell
8. © Lewis Jefferies / WWF-UK
9. © WWF / James Morgan
10. © Green_Renaissance
11. Credit to wellsie82 / Getty
12. © Jonathan Caramanus / Green Renaissance / WWF-UK
13. Credit to xPACIFICA / Getty
14. © Jeff Wilson
15. Credit to Ayzenstayn / Getty
16. © Jeff Wilson
17. © Studio Silverback
18. © James Morgan / WWF-UK
19. © Conor McDonnell
20. © Conor McDonnell
21. Credit to Jethuynh / Getty
22. © Studio Silverback
23. © Marizilda Cruppe / WWF-UK
24. © Studio Silverback
25. © Chris J Ratcliffe / WWF-UK

Contributor Acknowledgements

The introductions to chapters 4, 5, 6, 7 and 8 are © Sir David Attenborough, 2021, © Shakira Mebarak, 2021, © Naoko Yamazaki, 2021, © Christiana Figueres, 2021 and © Hindou Oumarou Ibrahim, respectively.

The following people quoted in the book were interviewed as part of the *Earthshot* TV series: Kip Ole Polos, Keiran Whitaker, Shulamit Levenberg, Cat Barton, Emma Seal, Henry Centeno Morada, Carmen Argueta, Neil Sims, Hiram Rosales Nanduca, Hugo Richardson, Deepak Mallya, Siobhan Anderson, Charles Massy, John McGeehan, Tom Szaky.

Every reasonable effort has been made to contact the copyright holders, but if there are any errors or omissions, John Murray Press will be pleased to insert the appropriate acknowledgement in any subsequent printing of this publication.